QH
309
S 3

This is our World

University of Oklahoma Press : Norman

This is our World Paul B. Sears

QH
309
S 3

Revised Edition

26612

Copyright 1937, 1971 by the University of Oklahoma Press, Publishing Division of the University. Manufactured in the United States of America. First printing of the new edition, February, 1971.

TO MARJORIE AND JOE
FOR THEIR WISE, UNTIRING, AND
AFFECTIONATE TYRANNY

Preface

THIRTY-THREE YEARS LATER

THE original edition of *This Is Our World* appeared in 1937, two years after *Deserts on the March*, which is still in print. In 1935 a twelve-inch shelf would have held the books on the imminent environmental crisis. Today, following Stewart Chase's *Rich Land, Poor Land*, William Vogt's *Road to Survival*, Aldo Leopold's *Sand County Almanac*, and Fairfield Osborn's *Our Plundered Planet*, books and articles on this topic make up a library of respectable size and quality.

Meanwhile environmental deterioration has become self-evident. Public opinion has been aroused and political action is responding. Population pressure, contamination of air and water, and urban decay are no longer matters of academic concern. Why then should the present volume—dated as it is in some ways yet dateless in others, serious yet somewhat whimsical—be reissued? The best answer, perhaps, is the reception of the earlier edition.

We (meaning publisher and author) have considered two choices: to reprint the original text with its anachronisms as a measure of the changes that have taken place in a third of a century or to touch up the more obvious of these without disturbing the essentials. Our judgment has favored the latter course, along with replacing the final chapter.

That the book has its shortcomings, no one can be more

aware than I, thanks to my own increased mileage, the rapid course of events, and the unprecedented growth of knowledge. But it does represent an early attempt to show that a sane adjustment of human society to the earth which sustains it is not primarily a matter of legislation or applied science. Rather it rests upon the invisible, elusive, yet all-powerful values and standards which, in any society, determine the ends toward which its instruments will be employed.

Actually *This Is Our World* was or is an informal experiment, an effort to combine, more or less colloquially, the wisdom to be derived from both cultural anthropology and ecology. Whether or not it did this, there is comfort in having been told by the late Clyde Kluckhohn that he considered it an important book for his students in anthropology at Harvard to read.

Something else that may interest the reader who keeps in mind the date of original publication is the discussion of issues that are lively today. These include not only the now familiar themes that are gathered under the rubrics of "ecology" and "conservation," but such matters as the insane destructiveness of war and the appeal of dictatorships to confused and wasteful societies. The warning clouds of these coupled tragedies were plainly visible in 1937, but even such a powerful voice as that of Winston Churchill went unheeded.

PAUL B. SEARS

Contents

	Thirty-three Years Later	vii
I.	The Living Symphony	3
II.	.... and the Modern Dissonance	10
	THE PATTERN OF INANIMATE NATURE	
III.	The Theater of Life	27
IV.	The Atmosphere at Work	41
V.	Water, The Great Sustainer	55
VI.	Soil, the Reservoir of Life	66
	THE PATTERN OF LIVING NATURE	
VII.	What is Life?	89
VIII.	The Pageant of Life	99
IX.	Plants, Life at Anchor	116
X.	Animals, Life on the Move	134
XI.	The Living Environment	155
XII.	The Pattern of Life	172
XIII.	Succession, the Growth of Communities	192
	THE PATTERN OF HUMAN CULTURES	
XIV.	Man, the Human Animal	213
XV.	Society, the Communities of Man	234
XVI.	Disease, the Failure of Adjustment	253
XVII.	Technology, Slave or Master?	266
XVIII.	The Challenge	281

This is our World

Chapter 1
THE LIVING SYMPHONY . . . .

WE are, literally, lucky to be alive. This earth may not be the best of all possible worlds, but it is the only planet known to us which can support such a system as life.

Mankind, in the mass as in the individual, has an immense egoism. Is not man the dominant organism upon the lonely, whirling earth? Are not the rocks, the trees, the animals all subject to us, here for our use? They cannot answer; they must obey. Should some trusted source of utility fail us, are there not others, ready from an inexhaustible earth at the touch of an equally inexhaustible human ingenuity?

For many thousands of years, with inconceivable slowness at first, but with gathering speed, man has spread the network of his culture over the earth. Brushing aside the confusions of history, we see that one culture after another has passed away, or else made peace with itself and nature by establishing some sort of balance. The price of wrong adjustment has been the same price exacted earlier from countless kinds of extinct animals and plants—oblivion.

Even where balance has been achieved, the advent of higher and more powerful cultures, covetous of the possessions of the simpler, has brought conflict, confusion, even extermination. This Nemesis unearned, this retribution for no crime is still in its course; it is the stinking basement whereon three modern centuries have built their civilization.

A courageous and beloved friend, the late Dr. Gayfree Ellison, spent most of his life in the healing and care of the human body. But in his youth he had been a soldier of the line, fighting in a wild and primitive foreign country. He once told me this story, as nearly as I can remember it:

"We were entrenched near a group of hostile tribesmen. We had set a line they would not be allowed to cross . . . . One morning early they appeared. On they came, big, brave, magnificent, naked men, with great shields made of the hide of the water buffalo, and long spears. We of course were armed with repeating rifles . . . . They crossed the line. We let 'em have it, mowed them down. *It was the saddest, most pitiful sight I have ever seen."*

Now it happens that the aftermath of victory was a wise

and humane colonial administration; and as I understand it, the tribe whose defeat is described became rather more attached to their conquerors than to their immediate neighbors. But that is not the point.

In one short paragraph we have a case-history out of the ceaseless conflict which has marked human movement over the earth. We have a view of the inevitable victory of technique and invention over courage with inferior equipment. And we have a brief glimpse of sympathy and the common humanity which is as much a part of the picture as is struggle.

Man is, after all, a part of mankind, and mankind a part of nature. Man is within, and not above, the vast symphony of environment and of all life. Man is, when all is said and done, a living animal. The noblest of his kind must bow before the laws which govern nature.

The demonstration of this fact, and of man's origin in a long and earthy past, has been rocking the world for more than a century—since the Darwinian earthquake. It was turned to immediate use for the good of mankind by the physician. Every doctor knows, or at least proceeds as if he knew, that man is an animal. His therapy, however, seldom calls on him to remind his patients of that fact.

In the field of daily life, this knowledge has served to refresh our memories concerning the wise words of Plato, who preached that to make the good man, you have first to be sure that he is a good animal. We have been a little timorous about following Plato all the way, and devoting the first years of schooling strictly to physical training.

Cousin Cramchild still holds the ferrule, but his grip is weakening.

And we have resurrected an older, cleaner, idea of modesty than we have had. Modesty is becoming less a matter

NOT JACKSTRAWS — BUT — THE TREE OF LIFE

of concealment and shame, than of having bodies well-cared for and beautiful, that need not fear the light of day.

Not only for the physician, but for all who must know the scientific realities which concern living organisms, the newer understanding of man's place in nature has been a touchstone. Mist and obscurity have been dissolved. The roll call of the living and the dead has ceased to be a game of jackstraws. It has become the vision of the tree of life, whose humblest twigs and noblest branches alike all lead back to its ancient trunk. Time has taken on new majesty and new meaning.

But any earthquake which clears away the hovels for a better city is likely to shatter some goodly buildings as well. For upwards of a thousand years the western world had been paving the way for modern science by a firm structure of belief in the reign of law. Its universe was an ordered universe, and the duties of man were laid down, clear, unequivocal.

The intimation that man is an animal became the signal to break ranks. If the red rule of nature was ruthless, so be it. A thousand years had seen the struggle between this and what had been proclaimed as higher law. It had even seen deeds of tooth and talon, of cruelty and revenge, done in the name of that higher law.

The higher law itself, and all of the achievements called spiritual, became naught but cunning animal devices, subtle means to hold the strong in check and keep the weak in line. The human mind was used to explain away as futile all that it, itself, had ever done. The All-Seeing Eye, like that of Polyphemus, was put out with a firebrand,

and morals became at best a matter of convenience; ethics became the rule that, whatever is, is right. Each living thing was in a hostile world, to eat or to be eaten, and had jolly well better make the most of it.

It is possible for a poison draught to contain its own antidote. For not quite fifty years, science, driving on to the bitter end with its assumptions, has been investigating

the behavior of living things under natural conditions, as well as in the laboratory. And if it has found little that will comfort the sentimentalists, it has uncovered much that will gladden the understanding heart.

It has found, to be brief, that Nature is indeed ruthless, but in her ruthlessness she plays no favorites. Justice is to be had, but it is neither for those who swagger, nor those who beg. It must be built, on Nature's terms, through understanding and painful toil. There is no event that is not numbered, that does not have its consequences. Conduct has become once more a matter of tremendous significance, continuity and destiny once more the rule of life.

This is not to say that any individual is free, in the old theological sense, and that the transaction is between him and Nature uncushioned. She works upon him, and he responds, through the medium of his culture. In this he lives, and moves, and has his being. It limits and compels him, true, at every step; yet it can be altered by his participation, his understanding, his will, and his

work. And within this elastic frame he is free to build toward peace with Nature, or his own defeat.

Instead of chaos and disintegration, the world of inanimate and living things is knit tight together by its common activity. And of that world, man is part.

Such, then, is our theme.

FOR WE ARE JOLLY GOOD FELLOWS!

Chapter 2

. . . . AND THE MODERN DISSONANCE

OUR feet hurt, so science puts us on wheels. We are weary from physical exertion, so it provides us with power and machines. We miss our absent friends, so it enables us to talk with them through space. We suffer from disease, so it heals us. Mention the slightest wish and indicate a willingness to pay for its granting, and almost literally, science grants it.

We live in Fairyland, a land of benevolent Djinn. But we are not a happy people. Trouble has merely shifted the garb in which it presents itself; the substance remains the same. We curse the carburetor instead of the flint nodule which refuses to flake delicately and evenly under the pressure of our tool of deer's horn. We mourn fewer dead warriors and stillborn infants, but more mangled bodies by the roadside. Hunger announces himself by means of a price list instead of making himself manifest in the wanderings of buffalo or the vagaries of schools of fish.

Trouble remains with us, however the world may change. There are not many forms of self-indulgence which can compare, for sheer luxury, with that of thinking about our personal difficulties. Lawyers, priests, doctors, fortune tellers, and friends, with a sympathy that is genuine, or even plausible in appearance, are always in demand. And very useful members of society they are, for there is no mental poison worse than the belief that one has a monopoly on hardship. The simplest and surest sanitation is the assurance that the listener understands because he, too, has his troubles.

No doubt of it, the plaints of other people, other places, and other times, seem strange to our ears if we hear them. The bluff old geologist, sharing potluck at many a campfire and sodhouse, has no patience with his epicurean friend who finds the Romaine and crepes Suzette at a restaurant not to his liking. The very recital ruins the geologist's day quite as thoroughly as the shortcomings of the chef render his friend's journey an unhappy one.

Yet from the caveman grumbling over a piece of stubborn flint to the housewife nursing her wrath against the plumber, or the ambassador in white fury over the seating at a state dinner, the essentials have changed very little. When the Greek comedy *Lysistrata* was revived after twenty centuries, it had all of the freshness of today. Suffragette and her warmer-blooded sisters, errant, ardent, and combative husbands—these and their troubles might have stepped in off the nearest street. It was nip-and-tuck for one seeing it some years since, which was the better show, the play itself, or the roaring enjoyment of the modern, midwestern, metropolitan audience.

Culture serves only to change the details whereby trouble expresses itself. The Romans forecast events by looking at the guts of a newly killed animal. Such visceral details convey nothing to us, but they meant a great deal to the ancients, on the eve of some enterprise. Yet the import of an editorial or a news item—nothing more tangible than the arrangement of symbols on a printed page— may mean the difference between hilarity and apoplexy to the modern business man. In either case, the root of the matter is the same. The gods have smiled or they have frowned.

Knowing these things, it is easy for the civilized man to be complacent over the troubles of his fellows, if not his own. Blues-singers are as old as mankind, doubtless much older. If the howl of the wolf is not melancholy, it is a first-rate illusion. Cassandras and Jeremiahs are honored by time, if not by their immediate audience. Every age has its pessimists, amateur or professional, but always vigorous. Man is born to trouble, so what of it? The philosophers themselves have had but three answers—forget it, avoid it, grin and bear it.

Nor is the soreness, the futility, and defeat merely a matter of the personal and trivial. It has soaked through the fabric of the whole group of mankind. Let us take the starkest, simplest basis for our judgment. What proportion of mankind has the opportunity, in exchange for a reasonable amount of exertion, to secure food, shelter, and clothing? We do not know, even in the United States, with all of its enlightenment, its means of gathering and communicating information, and its population as yet about one person for every ten acres of land. But we do know that the numbers who have not this opportunity are to be counted, not by the tens or hundreds of thousands, but by the millions.

Let us take, if we like, factors not quite so tangible, yet absolutely vital to a healthy mankind—peace and confidence. Where will we find them, within or without the nation? Where do we see a great leader arising to power through an appeal for common understanding, patience, and tolerance of others? How often are we reminded that we are all passengers[1] on the same planet, and that the injury of one means the suffering of all? Instead we see, from the smallest precinct on out, a consistent use of the ancient, cheap, easy, and brutal maxim—*Divide et Impere,* "Divide and Rule." Sow discord and suspicion among classes and groups, divert their attention to details, and

[1] Hendrik Willem Van Loon has brilliantly developed this thesis in his *Geography*. Simon and Schuster, New York, 1932.

power is the reward. In the language of the swimming hole: "Get the boys to fighting over a stick of candy, and you can steal their pants." It is a trick that is as musty as an old damp biscuit, yet it works as well as ever. It is the pushbutton of political life. But it is also a fatal device in a world hurtling on towards disaster as our world is today.

As within the commonwealths, so it is between great nations. One of our more candid public men is credited with having said: "There is no such thing as international good faith." Events seem to justify his skepticism with increasing force. Formal declarations of war involve embarrassing explanations, so they are discarded as a burglar discards the formality of knocking at the door in plain daylight. Victors do not sing as the Assyrian "of the young men's ears I made a heap; of the old men's heads I built a minaret," but modern warfare, with its disregard of neutrals and civilian populations, has returned to all of the savagery of the primitive world. Indeed to greater depths of ferocity and mechanized ruthlessness has it descended. Military men have a saying that the war between the states was the last gentleman's war the world has seen—Andersonville Prison, Reconstruction, and Missouri warfare included.

Not only are wars more ghastly. They are increasingly frequent, reminding one of the rapidly forming spiral vortex of a great prairie tornado. Moreover, they are consuming an increasing amount of the world's vitality. Industry and energy are being diverted to feed them. Of course there is growing sentiment against war. Much of it, however, is founded on nothing more exalted than the dying plaint of the philandering Johnnie — "Your bullets hurt me so." The basic curse of war is the fact that it destroys, not only civilization, but the world's store of energy which makes civilization possible. War no longer has the justification it had in a primitive and semi-civilized world, where it served to regulate populations and to diffuse culture. Western Europe has no occasion to be proud of her cultural contributions to home-turning armies.

To a greater extent than most of us realize we have come to accept the level, or even the direction of our industrial activity as normal when we use war or preparation for war as our yardstick. We labor that the means and fruits of labor may be destroyed.

There is no need to continue the recital of dissonance and evil much further. Every physician knows that disease wears two faces—the outward symptom and the inner disorder. Thus far we have been talking of symptoms. We could have named many more. We could, for example, have spoken of the commercial fabric of modern society, where money, the symbol, has become the reality. A means, an instrument, devised for the measure of realities, we mistake for the end in itself. In terms of energy and materials this is a fallacy as grave as to deal in pints and acres without specifying milk or land. The farmer, not sensing that money is but a measure, labors like a beaver to produce cash crops, neglecting his garden, his poultry yard, and his apiary. Thus he courts all of the vicissitudes of weather and shifting demands, in the end to buy a miserable vestige of what he might produce for himself.

Enough of symptoms. They crowd upon us at every turn and absorb our attention. We struggle to remedy them, for all the world like an adolescent, dabbing antiseptic upon pimples which will yield to nothing but a change of diet. Like his, our trouble lies deep within.

We conceive of our long and checkered past as a struggle of individual men and women to survive and adjust themselves. But the sombre truth is not so simple, so disconnected, and so episodic as that. The activities of man, the social animal, build themselves into a pattern. This pattern, as we shall see, governs his relations to his fellows and to the inexorable world of nature about him. We can carry on without bison or flint mines in these United States, for our pattern does not involve them greatly; likewise our predecessors would not have been greatly handicapped by the absence of oil, coal, or iron mines. So we wiped out the bison and ignored the flint just as the aborigine survived in blissful ignorance of those things which are the basis of our culture.

And so this pattern determines both how we shall get on with ourselves and how we shall get on with nature. Before the days of science there was no choice but to cut and fit, learning by hard and bitter experience. A few cultures, such as those of the Chinese, the Egyptians, the Indians, and apparently some of the minor democracies of modern Europe, achieved an admirable balance, if not security. Many others, particularly the more brilliant and energetic civilizations—almost the roll call of history—have succumbed, wearing out their welcome on the very soil which gave them their greatness. But because the means at hand were relatively simple, the end came slowly, at times almost imperceptibly.

The decline and fall of a culture is not a picture which Gustave Doré could engrave upon a single tormented plate. It is no struggling, twisted, mass of humanity falling into a vast abyss, no sudden wiping out of all that is.

DECLINE OF EMPIRE

Instead it is the molecular disintegration of a slowly rotting structure—a few individuals moving out or perishing here, a family or a clan degenerating there. Standards of living become lower, hope vanishes, suspended animation comes on the wings of spiritual discouragement and universal futility. You need not go back to the classic world to see it happening, if your vision is keen.

This, let us remember, is the danger that threatens any culture which grows in age and population. Just as a field, sown continuously to wheat, will deteriorate in the course of time while the natural forest or prairie beside it will be maintained in all of its richness, so it is likely to be with all but the most fortunate and well-balanced of civilizations. The immediate and visible cause of decline may be labelled military, moral, economic, or industrial failure. It may even be, as in the case of the wheat field, simply a matter of the depletion of certain soil minerals. But it represents, in reality, a failure of an artificial community to achieve the adjustment which enables the natural community to survive.

Recently a new and powerful element has entered the picture. Science, some of whose achievements were mentioned at the beginning of the chapter, holds out to us the promise of an end to the blind struggle of trial and error. Through her aid we are able to control the physical and living world about us, warping it to our own needs and desires. Here, it would seem, is release from any possibility of the fate which has menaced all civilization in the past.

But the trouble is that we have not so much listened to the scientist, as taken him in hand. We have used him, not as a consultant and friend, but as a handy man. He has been kept in the workshop as a trouble-shooter, not invited to sit with the board of directors. In fact, he has seldom been groomed for the latter and greater task. In his training, instead of being allowed the Grand Tour through the realms of the mind, he has been caught young and subjected to an intensely narrow and specialized apprenticeship. This has made him tremendously efficient within a narrow range, but has clouded his vision, his sympathy, and his knowledge of his own responsibilities.

Because the symptoms of wrong environment and maladjustment which impress us are the immediate and the personal, we have put the scientist to work on them. There has been plenty for him to do in the control of disease, production of food, structural materials, machinery, and fabrication of all sorts. The response has been immediate and startling. One generation has seen the length of average human life practically doubled. The dreams of Roger Bacon, Leonardo, Jules Verne, and H. G. Wells have not merely been realized—they have been exceeded beyond all dreaming. Not only do we enjoy mechanisms of the utmost ingenuity, but we have tapped the hidden sunshine of the earth—its stores of rich energy which have been built up in the long course of geological time. Man is, beyond all question or argument, the dominant species on the earth today.

But modern civilization, more than any which has gone before, is living visibly and dangerously beyond its means. The power and knowledge at our disposal are being used as never before, to drain the treasures of the past and borrow from the future. We are flowering for the sake of the immediate present, and at an accelerated speed. When pressed about the outcome, we merely shrug our shoulders and say "science will find a way." So rapid is the change, and so dazzling each passing moment, that it has come to seem as though the present were all-sufficient. The past, so dark, dim, and different can be forgotten; the future will take care of itself. Continuity and destiny can be left for the curious and idle. The busy world of affairs has neither need nor time to think of them. The Three Fates have been banished with all the other junk of our ignorant and fumbling fathers.

We have knit the peoples of the world together into a structure as seamless and as sensitive as a living body. With our greater power and speed the area of effect has increased. The old insulation of cultures which gave them an opportunity to develop into some sort of order, on a small scale, and without interference from others elsewhere, is gone.

Thus while the thing which governs our living is breaking up at the center, from within, through the application of science, it is breaking up at the edges through overlapping with other cultures over the earth. And so it happens that strong men in many places have stepped in to halt the chaos. Authoritarian governments try to create artificial islands of cultural unity and stability over the earth, to hold things steady in the midst of flux, hoping, quite apart from understandable and selfish motives, to develop some sort of pattern which will be in balance and survive.

We, on the other hand, are committed to a plan of democratic control—really a utopian notion, but magnificent and worth the effort. As compared with compact, centralized nations, we are awkward, sentimental, and slow-moving. We are like a bumbling fighter who telegraphs each blow seconds before he releases it. But we have unfathomable power and possibilities once we come to sense the realities among which we live.

We have chosen, in place of the old terse suggestions of

the philosopher when confronted with trouble—ignore, avoid, or endure—to put into the mouth of our new oracle, the scientist, this laconic formula "remove trouble." It has all of the challenge of an order of hope and of action. It has produced results, but at the cost of an upset world.

It is needful that we relax the pace at which we have driven this servant and guide, the man of science, so that he may catch his breath and speak for himself. Even his simplest services are based upon a principle we choose to disregard—that of *understanding*. Valuable as he has been in a practical rôle, his greatest service is that of an interpreter. We have not waited for him to picture the setting in which we must carry on, and to which we must fit. Soon we will have no choice—if indeed it will not be too late. Cleverness and facility must give way to wisdom and the long view. Once we have these we can let the details fit the larger perspective. We must forget the tinkering and look to the architecture.

We can even break up the framework of our activity into three segments for the convenience of our interpreter. Three phases of reality touch us at every turn, not always with the same directness, but all inescapably. The world of inanimate nature—the world of rocks and water, air and sunshine—is the first. Engineer and mariner, farmer and miner, work with them most directly. But the laws of their operation are inflexible for us all.

The world of living nature, of which we are a part, and from which we yet regard ourselves as curiously aloof, is the second. In its history and behavior, it is interwoven throughout with the inanimate world. And within itself, it is far more than the museum which we so often idly consider it to be. It is no mere assemblage of living plants and animals. It is a tightly integrated fabric, whose destiny is bound up with our own. The laws which govern its operation to a large extent are those which govern ours. Man's spiritual accomplishments will die of inanition if they are divorced from their animal basis.

And finally, the third of these three great segments of natural reality is the one which confronts us most immediately, often making us forget the other two. It is the culture pattern of mankind, now dissolving, and we hope, reforming under the impact of rapidly changing conditions. It puts the words in our mouths, the thoughts into our heads. It governs our actions, and those most intangible, delicate, and significant of our possessions—our feelings. Every experience which presents itself to us carries with it an unconscious appraisal of its value and significance. This is the work of the feelings. The matters of greatest importance to us—beauty and rightness—rest their verdict there. We say the feelings are innate and individual. Certainly they represent the one phase of human experience in which no consensus is possible, by any aid of science or reason. Yet they decide the greatest issues of living.

But if they are innate and individual, they too are a product in large part of the culture pattern, working in conjunction with the physical and living realities of the world. Certainly it is our task to know these realities, and the character of the pattern which brings them to an effective expression in human life.

So we shall take up the three great segments of the world of which we are a part in the order they have been named. And if at times the task seems tedious, we can be cheered with the thought that at the end, like the patient digger of fossil dinosaur bones, we can achieve some sort of resurrection in the assembly when we have finished.

THE AMISH
TRAVEL THUS

THIS IS THE
BY-PRODUCT

WE TRAVEL
THUS   THIS

THE BY-PRODUCT IS

FUNNY FELLOWS, THE AMISH, THEY HAVE
NO BREAD-LINES,
NO WORN-OUT FARMS!

THE PATTERN
OF INANIMATE NATURE

Chapter 3
THE THEATER OF LIFE

THE peculiar fitness of earth no doubt seems strange to anyone who knows how many difficulties life has met and become adjusted to, here upon the earth. The deepest, darkest abysses of ocean swarm with life. Red snow and hot geyser waters alike contain simple forms of plants which thrive nowhere else. Caves, deserts, hard rock faces, and even cakes of solid salt afford means of existence to appropriate organisms. Earth is varied and resourceful, to be sure, but there are limits beyond which it cannot pass.

A fussy little old lady, involved in some personal difficulties, went to an attorney for advice. The lawyer, a kindly and conscientious man, explained to her the merits —and demerits—of her case. Indignantly she arose in protest, "That can't be right! The neighbors told me just what the law was."

Many of us get our science as the lady preferred to get her legal advice. Writers of romance, and those modern counterparts of the Pied Piper—comic-strip artists—having exhausted the humdrum earth as a source of interest, are becoming astronomical, always with enthusiasm, sometimes cleverly. Thanks to their efforts, most of us have a picture of space teeming with planets or other bodies which either bear some form of life, or are swept and garnished, ready for that great adventure. So far as other solar systems than our own are concerned, no one can gainsay the fact that they might be right. But concerning the planets which are the sisters of Earth we, in their own language, "have the goods."

For one thing, even the simplest form of life is, after all, an extremely delicate piece of mechanism. And anyone knows that the more delicate an instrument, the more exacting are its requirements for smooth operation. You can use the same pair of tongs to handle cold ice and hot embers. But you could scarcely expect your watch, adjusted as it is for temperature changes, to keep the same accurate time in a freezing tray or in the steam of a teakettle that it does in your pocket.

Like the watch, life has temperature limits beyond which it does not run smoothly. True, we marvel at the way typhoid bacteria will live in a block of ice and seeds awaken to vigor after a long winter's exposure. Animal eggs and resting stages often withstand zero temperatures. The deadly botulinus or food poisoning organism will live in spite of canning methods that kill ordinary bacteria. But there are still further limits of heat and cold beyond which even these various hardy forms cannot remain alive.

When we turn to that most intricate expression of life known as human civilization, temperature is a very important matter indeed. Men cannot do their finest and

most creative work if they are too hot or too cold. Some say that such work is impossible where the temperature, no matter how agreeable, does not change rather violently at times; but the sages and artificers of Egypt or India might not concur in this.

The mechanism of life is inseparably bound up with water. Thirst, either in plant or animal, is the dread herald of death. Not only do we know the need of water from common experience, but that wizard of the modern world, the physical chemist, assures us from his calculations that only one known substance might conceivably do the work which water does for such an energy system as life. That substance, curiously enough, is ammonia, and it is out of the question for several reasons.

Among many other uses, water is necessary to bodily form, to internal circulation, and to movement in living things. The parts of an animal move about in a bath of water, much as the bearings of a car move in a bath of oil. It is clear that to be of any use for these purposes, water must be neither in a vapor nor in a solid form, but in the liquid condition. And so the limits of ordinary activity lie between the freezing and boiling points of water.

As a matter of fact, the limits are generally narrower than that. The living stuff of plants and animals—protoplasm is its name—contains a good deal of protein of one sort or another. Now gelatin and albumen are proteins, and anyone who has made a gelatin dessert, or watched the poaching of an egg knows that these substances are very sensitive indeed to temperature change. It does happen that the protoplasms of different organisms are quite varied, so that one's heat is another's cold. But as we look them all over, it is difficult to find many forms which can endure the temperature range which is, with the aid of clothing, fire, and ice, tolerable to man.

Let us continue with the engineering specifications for life. In addition to water and suitable temperature there must be an atmosphere—an ocean of gases in which life can be carried on. Many of the most important activities of plants and animals depend upon the intake and output of matter in the gaseous condition. Anyone knows that a candle will not burn without air; and air is quite as necessary if the body is to perform work, for much the same reason that it is required by the candle. Moreover, the air is a great reservoir of that most essential and remarkable constituent of life, the element carbon; also of nitrogen which is found in all proteins. From this great reservoir, as from the minerals of the soils, life continually withdraws what it needs. Continuously, too, it returns materials in the form of waste, to be used by future generations of plants and animals. In short, no form of activity even remotely resembling life is conceivable without an atmosphere.

But life exacts even more from environment. It must have a continuous supply of suitable energy. Life itself is in fact a self-perpetuating manifestation of energy. This energy it can transform in the most amazing ways. Yet it cannot, by any means we know, create that energy. Energy can be obtained in many ways, but earth possesses a considerable and constant source in the radiation which it receives from the sun. And it is this solar energy, largely the visible or luminous portion of it, which makes the wheels of life go around. So long as the sun remains hotter than the earth, which is likely to be for a considerable time, by our standards, so long will the flow of energy continue. We shall see later how this energy is fixed by the activities of green plants and thus made available, not only for their own use, but for all of the activities of the animal kingdom.

It is this same fixed or stored sunshine, altered into oil or coal which plays so large a part in the modern industrial world. No one who is familiar with the elaborate means for transforming and using it would accuse us of failure to understand its importance. But it is well to remember that civilization is coasting along upon the accumulation of ages without much thought of what will happen when the surplus begins to run low. It is also well to remember that, once run through the mill of life and reduced to the form of heat, the sun's energy can serve but one purpose—that of raising the temperature of the earth.

It is forever, so far as we know, lost to the use of living things. And in passing, too, we might remind ourselves that there is absolutely nothing that compares with a first-class modern war to speed up this melancholy process.

This is no mere idle humor for the laboratory; it is grim business for the statesman.

Now that we have seen some of the prime requirements of life, let us have a look at the earth once more. In the first place, the earth is at a happy distance from the sun to meet the needs of living things. If it were too close or too remote, it would be too hot or too cold; its water would be vaporized or sealed into eternal ice. We have seen that either condition would make life impossible. And we might add that this applies to upside-down giants and blue trees and anything else that can be imagined in the way of living creature.

The size of earth, no less than its distance from the sun, makes it the suitable abode of life. Were it much smaller the force of gravity would not hold an atmosphere. Like the moon, it would be not only cloudless, but airless. And even if life could dispense with an atmosphere, which it cannot very well do, this too small earth would be a dry mass of rock, for its water would be lost as fast as it turned to vapor. Day would be a time of torrid heat, night a season of intolerable cold, since water is the great regulator of temperature.

Were earth a great deal larger than it is, the force of its gravitational pull would hold an atmosphere so dense that, if the essential water were present, clouds would obscure the sun and cut down the effective radiation which, as we have seen, keeps life going.

And finally, the earth might have an atmosphere of quite the wrong sort to support life. It is conceivable that life might get the carbon it needs from mineral sources, instead of from the carbon dioxide of the air. But this carbon dioxide is quite as remarkable and necessary a substance in its way as water. It not only forms very useful

combination with water, but it acts as a great chemical stabilizer in the environment, precisely as water acts to stabilize the temperature.

We hear much of the survival of the fittest. Certainly a species will be wiped out if it cannot adjust itself to the conditions under which it must live. But to regard this principle as the basis of all biological thinking is as dangerous as to build a science of economics on the assumption of universal avarice, or a legal system on the principle that every citizen is a criminal. Many features survive the test of time which have little or nothing to do with fitness; and the measure of fitness itself changes with time. Louis the Fourteenth or Henry the Eighth might have some difficulty in adjusting their unquestioned talents for questionable enterprise to the modern world, at least in their old rôles.

Quite as important as fitness of the organism is the fitness of the environment. And that is why we have gone to some length to explain the peculiar fitness of this planet, Earth, as the theater of life. We are, as we have said, lucky to be alive. We know, through the marvels of the spectroscope, that the laws of physics and chemistry which apply here on earth extend in their operation to the limits of the visible universe. And we know that it is a matter of rare chance that gives our earth the combination of favorable conditions which she possesses. We have been taught to respect our bodies and our talents, to keep them fit for the conditions under which we live; but we have given too little thought to the corollary. The environment, too, must be respected and kept fit, so far as it lies within our keeping.

Of course we have little enough to do with affecting the great building stones of environment which have just been discussed. But we and all other living things have a great deal to do with the pattern and relationships which they assume, as will be seen in the course of succeeding chapters. Meanwhile, let us examine their pattern somewhat more closely.

The earth, a spheroid, spins as it moves slowly about the sun in an elliptical pathway. Sunlight falls upon it in a broad band, as a wide brush might paint a stripe upon a whirling, globular bowl. Changing position and distance bring the rhythm of night and day and the succession of the seasons with a regularity which is the model of all terrestrial clockwork. Moreover, the pattern of sunshine and shadow has a regularity which is the beginning of all geometry. Yet when we examine the results in terms of temperature and humidity patterns over the earth, everything seems strangely warped. Lines which mean so much to the navigator and surveyor are of only incidental interest to the student of environment.

This is because the eternal rocks, as we like to call them, are really a very much mixed-up and unstable assortment, forever crowding themselves about, creating pressure here and yielding there. And so it happens that there are the towering Himalayas as well as depths of ocean into which those mighty mountains might be dumped without a single peak reaching the surface. Earth's water, which might serve to cover the whole surface for a little depth, is in these depressions for the most part, while the land surface too is in great masses, the continents.

Thus the broad band of sunlight, played so regularly about the spinning earth each day, falls upon a very irregular surface. Land heats and cools far more rapidly than water, thus suffering more abrupt change between day and night, winter and summer. Water, under the influence of sunlight, becomes vaporized, thus charging the air with moisture. And so we have at once two great types of climate to break the evenness of the pattern of sunshine.

The oceanic climate is marked by great humidity and modulated seasonal change; the continental climate by dryness and abrupt change.

We might expect to find the most extreme type of continental climate in the precise center of each great landmass, shading away in all directions into the oceanic at the margins. Instead of that it is displaced, by the whirling of the earth and the greater average heat of the tropics, toward the western equatorial portion of each continent. We have but to recall the Algerian, Andean, Californian, and Australian deserts to appreciate this fact.

In addition to the oceans of water with their profound effect upon environment, there is the ocean of air some miles deep, at the bottom of which we live. Within these two great fluid systems the energy of the sun sets up currents and mass movements. Like the water in a saucepan, or the air above a bonfire, fluids expand and become lighter when they are heated. Thus they rise and colder masses rush in to replace them. Ocean and air are ceaselessly turbulent, as we have seen the great rock masses to be mixed and unstable. Small wonder that life consists of endless unrest and activity, for it must be lived in a world which knows not repose.

There is a measure of regularity to the currents and the winds, due to the form and motion of the earth. But it is sadly broken and thwarted by the very irregular pattern of land and water. Yet with all of that, it determines the underlying pattern of life and of human activity. So delicate is the relationship that we can test the validity of our calculations and measurements of climate by looking at the vegetation; far better in fact than we can by parallels of latitude and lines of longitude.

The moving masses of air, which are the result of temperature change, are also a means of producing it. Moreover, they are the means by which moisture taken from the sea is moved about and distributed. This moisture, falling upon the land masses, soaks in where it can, gathers in depressions, flows away, or evaporates where it cannot. The highest points upon which it falls are the mountain tops of the thin upper air, exposed directly to the remorseless forces of sun, frost, and wind. Not even flint or porphyry can endure such exposure without cracking and crumbling. And the water, as it falls and flows away, grinds and washes this loosened material with it, down into the lowlands and on out into the sea. Thus are the continents worn down like the teeth of an old horse with the passing of time. Eventually, however, the shifting of materials sets up new stresses and strains in the earth's crust; then new continents are born, new seas take shape. In terms of human life, of course, such changes are extremely slow—a slight submergence here, a few inches of elevation there. An occasional landscape stripped bare by flood, or a city shattered by earthquake serves, however, to remind us of the titanic forces in whose play we are caught.

Into such an earth as we have briefly sketched, life has come. Launched we know not where, but nurtured in the sea, then cautiously it fared forth upon the land. There, upon and within the crumbled fragments of rock it lived and died for generations, until a new thing under the sun —the soil—came to be. And from that moment on, the work of living things began to build itself into the environment. Soil is not only the means of life; it is the record and result of life, interplaying with the rocks, the sun, the winds and waters. By its enrichment life has been enriched. Soil is a very precious thing indeed, the dark mantle of the living earth, as we shall see.

As the soil grew, it was guarded and husbanded against the elements, in a measure, by the plants and animals within and above it. This living cover tempered the force of climate for it and for each other. Thus increasingly did life become a part of its own environment.

Moreover, as we shall see, the fabric tightened as time went on. Organisms arose which depended upon each other, or upon groups of other organisms in many ways. And the newest of these in any significant sense, at once the most powerful and most dependent, is man. He is most powerful because he is fittest for the world as he found it; most dependent, because he is not fitted for that world as it had ever been before. With all of his resources he could not find place in the raw environment of the primal elements. He must live in a world where those elements have been infinitely tempered by his living environment.

Let him destroy that living environment, or change it too greatly, and he will find himself turning the clock, not forward as he may think, but backward. The priceless soil and all that it produces, will be swept down to sea even as the crumbling stones of the naked mountain tops are washed away.

Chapter 4

THE ATMOSPHERE AT WORK

THE narcotic of civilization makes us forget many of the realities of the natural world. But with all of our snugness and ingenuity, we remain acutely conscious of the power of climate and its tangible expression, the weather. Climate still shapes the broad pattern of human activity over the earth, while weather may turn the tide of battle and affect, not only the comfort, but the fortunes of mankind. However much we may plague and ridicule the efforts of the weather bureau, its prognostications are standard front page news, the first item to be scanned by millions of readers.

Society is kept going by the willingness of men to take risks, and recognizes this fact with a rough sort of justice in the appointment of reward—if nothing is ventured, little is to be gained. Of these risks, weather remains one of the gravest and most frequent. The loss and disappointment of a single farmer does not stop at his boundary fence. Multiplied by the score, the hundred, or the thousand, it carries on into the world of commerce and must be absorbed there.

A wholesaler whose business is the handling of perishable produce was explaining to a friend his habit of making wagers at the slightest opportunity: "This kind of

BOX CARS!

gambling rests me. I couldn't stay in business if I did not gamble by the carload every day."

A large shipment of lemons, arriving during an unseasonably cool week of the summer, might have to be sacrificed. Nor are conditions at the point of arrival the only source of concern. A few days of the wrong kind of growing or ripening weather, and a trusted source of supply might prove worthless. And all of the facilities of modern shipment and storage may sometimes be taxed by unfavorable conditions en route.

As weather affects the details of our enterprise, so climate governs its pattern. As the world becomes our unit, design of housing and clothing and the introduction of new crop plants is governed by similarities of climate. Only by virtue of new methods of transportation and preservation has the exchange of perishables far across climatic boundaries become a fact instead of a fanciful dream. In fact, the whole tapestry of human activity takes its varied theme largely from the variation of climate over the earth. In the long run, it must do so; but often, before the adjustment is made, there are many costly attempts and failures.

To obviate these attempts and failures through a better understanding of climate would seem to be the civilized approach. But here, as at so many turns in our complicated existence, the immediately practical interferes. We must, if possible, know the weather which is at hand. Most of the machinery for studying it is burdened with this task of short-range prediction. This knowledge is vital to the mariner, the farmer, and the merchant. Moreover there is no possibility of studying climatic patterns without long-continued, detailed daily records. The best we have are but a few-score years in length. Conscientiously as they have been made and kept, there are still many serious gaps in our knowledge; evaporation has proved itself a stubborn factor to estimate, for example. Yet in large degree it determines the effectiveness of rainfall, and the responses of plant and animal life.

Even with the best of records, the vagaries of weather are so great, and seemingly so capricious, that to analyze the pattern behind them is extremely difficult. From 1933 to 1937, continental United States suffered from severe drought, with unusually high temperatures. While searing sun and dessicating winds brought despair to the American midlands, rainfall of extraordinary amount plagued much of continental Europe. So great was it in Spain during the winter of 1935-36 that the weather may justly be counted as one of the immediate causes of the Spanish civil war; while in the United States, the climate certainly modified the course of politics, just as the favorable weather of the 1920's helped to create a disastrous optimism.

At the height of the drought in the United States, the Shawnee Indians of Oklahoma held a rain dance. Following ancient ritual, they used firewood from the dry creek beds to make the ceremony potent. They augmented their prayers for rain by harnessing turtles in these creek beds at the level where the water ought to be but was not. The night following there came thunderstorms, reviving the parched corn and freshening the grass, relieving the deadly temperatures of more than 100 degrees.

True, the harassed weather bureau officials, on the basis of telegraphic reports of air pressure, temperature, wind movement, and humidity from stations scattered all over the country, had predicted rain about as far in advance as the ceremonies were arranged. By and large, however, the Indian rainmakers have an excellent record, which probably means that they take few chances. The weather man is obliged to take more chances and make more errors.

There are white men who never enter a church and who scorn the weakness of their pious neighbors, who yet speak with awe of the power of the Indian rainmaker. Their inference is that the Indian must be closer to the ruling powers of the universe than our own black-coated clerics. He may be, at that, but so far as weather is concerned, sober judgment insists that his success is based more upon knowing when to pray than upon knowing how. And in tribal ritual, let us not forget, the rainmaker is traditionally the sole judge of the opportune time for his ceremonies.

Such ceremonies must be almost as old as the race, proof of the timeless hold and importance that weather has in the history of mankind. It is a grave mistake to believe that because primitive man had no recording, graduated instruments he had no means of measuring atmospheric behavior. The leaves of a tree are at once vane and gauge of the lateral movements of air currents. Their ascent and descent is likewise revealed to the child of nature by the smoke of his campfire. The pattern of sunshine and cloud he can see for himself as well as though he were the graduate of some great technical school. By their changes in texture, the rawhide thong of his bow or the twigs and leaves beneath his feet advise him of changing humidity.

For temperature he is by no means obliged to rely solely upon the nerves of his own skin; the pitch of the locust's rasp and many features in the behavior of various animals serve to verify his own impressions. The physical activity of reptiles and of many rodents, for example, can proceed only within rather definite limits of temperature.

Add the Indian's prolonged tribal experience with the functioning of natural laws as a vantage over the brevity of our own continuous weather record—less than a century—and it can be seen that the cards are not completely stacked against our red brother. The time may even come when we will have occasion to regret that the exchange of information between Indian and white, like the exchange of bullets, has been so poorly balanced.

While we are at it, let us work the good fellow for all he is worth. If he was weatherwise, he was also in tune with climate, and by no means unconsciously. The proof of that may seem obscure to some, but it is perfectly sound. He turned over to us a continent whose wild life and soil bore none of the scars of exploitation. He knew (and we can be quite certain about this) that lean years will surely follow the fat, and so he was never really wasteful of the bounty which lay at his hand. We must, of course, admit that he had limited technical means at his disposal; but there are cases where human cultures have killed the magic goose with means quite as simple as his.

The Indian had few personal possessions and no conception of private profit. He ate when he was hungry. He respected the plants and animals which furnished his needs; the thought of their wanton destruction was abhorrent to him. In time of plenty he made provision.

By contrast, on the very plains and prairies where he had survived good years and bad, our own bankers and farmers accepted the fat seasons as their standard of action and value. Plans, prices, and loans were made on that basis, instead of on the basis of a reasonable average. In con-

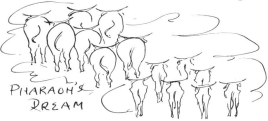

sequence the inevitable lean seasons could not be taken in stride; when they came, the whole mechanism of life was stunned and paralyzed. One experience with hobble or whip is enough for a reasonably smart horse. But we have been trapped in the mesh of climate at least three times, once during the 1870's, again in the early 1890's, and again in the 1930's. What price civilization, indeed?

The intricate pattern of climate and its transient expression, the weather, are due, as we have seen in the preceding chapter, to disturbance arising when the sun's energy strikes our extremely varied earth. Within the regular framework provided by the march of the seasons are all sorts of irregularities occasioned by the arrangement and behavior of land, water, and air. We have seen that the currents of the atmosphere, the shiftings of its great air masses, distribute moisture and rearrange the smooth course of temperature. We have considered the great division of climate into its oceanic and continental types.

In determining the climatic pattern of an area we are forced to consider not only the averages of temperature, rainfall, humidity, wind movement, and sunshine, but also their distribution through days, seasons, and years. Forty inches of rainfall confined to the growing season will have a very different effect from the same amount in winter, or evenly distributed throughout the year. Nor can one factor be considered apart from the rest; we have already suggested this with regard to rainfall and evaporation. A given amount of rain is far more effective in a cool or temperate climate than in the tropics. Likewise, much less energy reaches the earth where the air above it is heavily charged with moisture or dust than where it is dry and clear. The light of mountain tops is richer in ultra-violet than that of the lowlands.

Yet to discern the existence of a regular pattern in all of this complexity is not impossible. Brief as our records are, they show a sort of throbbing recurrence of groups of similar years. There is even enough of crude regularity in this recurrence to justify some in speaking of climatic cycles, although it must be remembered that the word "cycle" is a very sacred one in mathematics. These fluctuations have been pinned down to intervals of 6.5, 11, and 22 years. Even so, they must be thought of as trends, rather than events of precision. Since the records began in the mid-continent area, each year has seen certain places suffering from drought, other places with an abundance of rain. Recurrence rather than regularity is the thing with which to reckon.

We are by no means bound to our own written record of climate. Such wholesale changes as those of the glacial period have left their record in the very landscape. Changes of a lesser, but still millenial, order have been recorded in the deposits of ancient lakes. There the sediments have piled in order of receipt, like the letters of a snowed-under professor in his file basket; and these sediments tell a plain story of great climatic changes. Studying them, Professor G. Erdtmann of Stockholm is moved to think that we are living in a period of interglacial climate, and my own studies in North America seem to confirm this. But no one need take alarm; such changes are extremely slow. Their coming heralds its approach so that living things, even human culture, make the adjustment without conscious difficulty.

Changes in terms of centuries, as well as year by year, are also recorded for us most sensitively in certain tree rings. So delicate, yet definite, is the pattern, that the record of living trees can be tied on to that of logs and beams cut centuries ago and the story pushed back by that means. In this manner Professor A. E. Douglass has been able to construct a precise climatic history for the Southwest, and even to date quite exactly a number of prehistoric structures. Interesting and important for its own sake alone, such work is of priceless importance to us if we are to control in any measure our own destiny. As vividly as Pharaoh's dream it warns us of the inevitable fact—*climate fluctuates, lean years follow the good*. In comparison with that rule, our feverish legislation to safeguard business and commerce is as the play of children.

Suppose, lest we be thought to deal in the glittering general, that we examine some of the actual effects of climate upon living things.

We have spoken of good and bad years. These, of course, are relative terms. A series of decades of cool weather would push the northern limit for cotton and corn southward; but it would also extend the wheat belt farther south than it is now. Similarly, prolonged dryness would kill many maples and beech trees in the eastern states. At the same time it would favor the growth of oaks and prairie grasses. In the prairie itself, the tall grasses would suffer, but short grass, cactus and yucca would move in to replace them. The net effect, however, of dryness would be to decrease the total production of plant material. Since this is the measure of animal life, dry years are, for us, usually bad years.

Temperature is somewhat more complicated. In general, if there is an abundance of moisture, warmth promotes plant growth. On the other hand, many plants and animals whose products are highly desirable cannot thrive at high temperatures. Potatoes and rye are good examples of this. Quite aside from the question of human comfort, there are advantages to be had in cool regions, thus offsetting their somewhat lessened production. Even so, until low temperatures are reached, the reduction of material is not as marked as that produced by decreasing the moisture supply.

The effects of temperature vary with water content. Seeds and bulbs will endure heat or cold more safely if they are well dried, as a rule. In animals, the temperature requirements may even change with age. The delicate

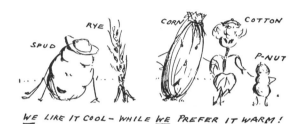

WE LIKE IT COOL — WHILE WE PREFER IT WARM!

corn borer larva cannot stand even a moderate amount of heat and dryness, although the mature moth might be insensible to them. This fact, by the way, safeguards the western corn belt from the scourge which has been so disastrous in the humid East.

Seaweeds are quite sensitive to temperature, rarely having a wider range for any species than about 10 degrees mean annual difference. The brook trout will not thrive in water that is not cold. Seeds of rye will germinate upon a block of ice, although they prefer greater warmth for further growth. Man manages to get around the problem of temperature largely through invention and skill and has thus become a highly cosmopolitan species.

We can get a fair idea of the relative importance of moisture and temperature by observing that the three great belts—forest, prairie, and plains—of eastern United States run north and south. They run across temperature lines, and with the rainfall. Moisture thus determines their form, and the amount of material produced. If however, we proceed along any of these belts from south to north, we will find some species dropping out and others taking their place. Temperature has assorted the kinds of trees or grasses, as we have seen that it assorts the kinds of crops we grow.

Light is another factor in climate. Like heat, it comes to the earth as radiant energy from the sun. It enables plants to make food for all living things. But just as the most perfect engines made by man can salvage only a fraction of the energy present in fuel, so plants utilize but a small fraction of the radiance from the sun. Small though it be, nevertheless this fraction does the work of life.

Plants, and animals, too, differ widely in the amount of light which they must have. Moths fly by night as a rule, butterflies by day. Weeds thrive in the powerful sunshine, while the beech and hemlock are injured by it. Strong light is injurious to living matter in any form, if the exposure is too long—hence the value of light as a sterilizer.

Light not only varies in intensity, but in quality and pattern, with resultant effects upon life. The ultra-violet rays, source of sunburn to us, affect the proportions of plants and take part in the formation of the very important vitamins. Much of the seasonal rhythm of growth and flowering is dependent upon the relative length of day and night.

With all of its importance, the utility of light remains a relative matter. The desert with its perpetual sunshine may become the food factory of mankind if it learns the secret of harnessing this energy. As it is, plants stand far apart in the desert, unable to utilize the energy which is there, for lack of water. Very often we accuse our trees wrongly of shading out the lawn beneath them; death may be due to lack of moisture.

One of the most serious problems created by that monster, the modern city, is the dearth of good sunlight.

Wind, besides the part it plays in distributing moisture, has its effect upon life, too. It promotes evaporation, with good results or ill depending upon circumstances. The farmer of the corn belt dreads the hot winds of midsum-

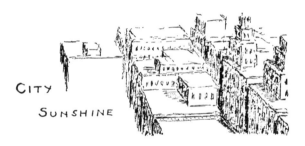

CITY SUNSHINE

mer. For they may evaporate water from his crops faster than the parched ground can supply it. Wind also assists in the spread of many plants, and even animals. It dwarfs the trees at timberline into grotesque shapes, sucks the dry soil from beneath our feet, and enables the waves of ocean to batter and carve the edges of continents. Like rain, it is a powerful force of change.

We have by no means completed the roll call, but enough has been said to explain the very direct operation of the atmosphere upon the living.

WIND-TIMBER

Chapter 5

WATER, THE GREAT SUSTAINER

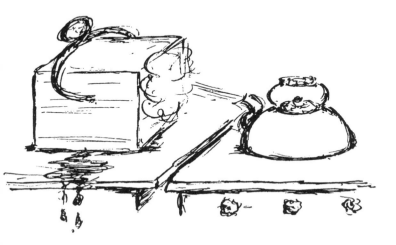

EVEN the sodden inebriate who claims that water, as a beverage, is a miserable failure, gets no hearing unless he first recites a long and impressive string of its virtues. We use water, we are surrounded by it, indeed to a large extent, we *are* water.

Withhold water from any living organism and it, of course, will die. But suppose our curiosity moves us to continue the experiment beyond that point; suppose we dry out the remains quite completely. If we do, they will lose anywhere from two-tenths to nine-tenths of their weight. Still we are not done. Suppose next we call in a chemist and ask him to remove the water which is chemically held in the dried residue. We would find that this represents a fair proportion of what is left.

If our taste does not run to the gruesome, we might take a map of rainfall and get it well in mind. Then let us take a map showing the price of an acre of good average farm land. They will vary in many respects, of course, but in general pattern both maps will agree quite satisfactorily. Or we might take instead, for comparison, a map showing the average size of farm. If we do, we shall find that it fits our rainfall map much like the developed film

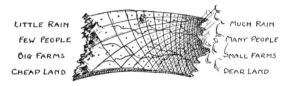

fits the photographic print made from it. The greater the rainfall, the smaller is the farm required to support an average family.

Or we can, if we choose, compare our rainfall map with one of population. Again we shall find, in spite of some interesting differences, a correspondence. The New England states, with upwards of forty-five inches of rainfall a year, will show several hundred people to the square mile. Until leisure, wealth, and technology changed the picture, parts of the arid Southwest, with less than twelve inches, were sparsely peopled. And the population in between will shade away much as the rainfall does. As for exceptions, there are many, of course. Where the air is very humid, a little rainfall does more good than where it is very dry. Unusually favorable centers for commerce, manufacture, and mining may produce so much wealth that water and food can be brought from considerable distance to offset a meager rainfall. And of course there are great river valleys in which a dense population is supplied with water by rain which falls perhaps hundreds of miles upstream.

If you like vivid contrast to bring home the truth, visit the front range of the Rocky Mountains. Admire, as you cannot help doing, the rich green turf in the front yards of Laramie, Denver, and Pueblo. Then hunt up some vacant, unirrigated lot; it will be a rectangle of sparse, tough, dry, semi-desert vegetation. Or drive eastward across the narrow belt of beautiful irrigated cereal, fruit, and vegetable farms which run along the mountain front. It is as lush as Maryland at her best. But it stops sharply where the water stops, and at its eastern limit you can step across a barbed wire fence into the arid gray of the High Plains.

Should your mind run to figures, consider this. Some time ago Professor J. E. Weaver studied the amount of wheat produced from the same area at Lincoln, Nebraska; Phillipsburg, Kansas; and Burlington, Colorado. Temperature conditions are about equally favorable at all three places, but the rainfall is respectively 28, 23, and 17 inches a year. The corresponding yields were 603, 402, and 353 grams in 1921, and 442, 316, and 224 in 1922 for the same unit area. A little simple division will show that these yields are not in close ratio to the average rainfall.

However, Dr. C. Warren Thornthwaite later developed a means of calculating the efficiency of rainfall by taking into account the loss produced by evaporation. His corrected figures for the effective moisture in each of the three places are as follows: 51.4, 38.0, and 30.0. Even if figures are rather a dull matter for you, it might be interesting to divide these numbers into the respective figures for yield, both for 1921 and 1922.

Try as we like, we cannot get around the fact that water is one of the most significant measures of the abundance and variety of life. The authorities in more than one great American city are puzzled to find new sources of water for their extravagant needs. Should they fail, their cities will simply have to stop growing. Some cities use well over one hundred gallons per capita each day, although a toughened campaigner can get along with a couple of quarts, if he has to, and still not be filthy. Of course not all of this demand is due to modern plumbing, by any means; industry is a greedy user of water, and many department stores and great hotels have found it necessary to have their own wells.

So pressing is the need, in fact, that the same water must be used repeatedly. The ancient Law of the River provides that every dweller along its course has a right to its water, quality unimpaired, and quantity not seriously diminished by those who live upstream. This law still holds, but it is not often earnestly invoked. The great city of

Cincinnati, for example, seems to find it more expedient to take the Ohio River as it comes, bearing the industrial and domestic wastes from above, and then purify the water, physically, chemically, and biologically for its own use. When one looks at the river today he finds it hard to realize that the old flatboat travelers used to throw buckets overboard and drink the pure water of that stream. But Cincinnati really has an excellent water supply and is justly proud of that fact.

Water, suitably governed and controlled, is all to the good. We have seen, and shall see further, that in nature undisturbed by man, its behavior is regulated by soil and vegetation. Not all of it flows down to sea, or evaporates into the air, by any means. The earth, both in the soil and in the rocky recesses beneath, becomes a reservoir upon

EVAPORATION — IMPORTANT BUT HARD TO MEASURE

which life can draw. Springs abound, and streams on the whole, move along with a fairly uniform flow. And the richness of primeval nature is made possible by this fact.

No one needs to be reminded that water, ungoverned, is a curse. Too often, however, our understanding of that curse is based upon headline material, dramatic and sensational, rather than upon the dull reports wherein lies the truth. When water ungoverned wipes out or paralyzes a great city, we wheel out the engines of relief, repair, and prevention instanter. Nothing, it seems, could be grimmer, more serious than the immediate devastation itself.

But there is something grimmer and more serious behind it. For repeated, ungovernable flood is a symptom of destruction, not merely for the works of man, but for all that makes them possible. High water, of course, is a perfectly normal, recurrent event in nature. The structure of the river valley, and the presence of its flood plain, are eloquent proof of that fact. Whoever builds or farms within that plain must know that he runs a hazard and plan accordingly. The resources of the modern engineer are equal to that challenge. When the swollen Potomac began to lap about the edges of Washington's public buildings, it became obvious that their basements were a very poor place indeed to store precious archives. Before the days of levees, valley farms in eastern Arkansas might lose an occasional corn crop through high water, but the gentle rise brought rich alluvium, more than enough to make good the loss. Today the levees take care of ordinary high water, but if the levee, under extraordinary burden, should break, everything would sweep away.

As a matter of fact, the recurring, devastating floods of today are much more than the normal springtime swelling of our streams. As they have increased, the flow of these streams has become less regular, and the stagnancy or dry beds of midsummer steadily more marked. Springs and wells have dried up, and the general level of the great underground reservoir upon which we all depend has dropped. Nor is this lowering a trivial matter. In eastern Kansas and Nebraska it amounts to some fifteen feet, at least. Farther east it is in many places more than fifty, and the water now secured by drilling wells to these lower depths is much more apt to be salty or otherwise poor in quality.

To put the case with brutal frankness, the continent below our feet is slowly drying out. The works of man all conspire to get water off the ground and into the streamcourses as fast as possible. Everywhere we have substituted for the old, leisurely natural drainage pattern an entirely new system—our highways. Only in a few places,

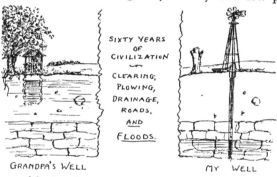

GRANDPA'S WELL

SIXTY YEARS
OF
CIVILIZATION

CLEARING,
PLOWING,
DRAINAGE,
ROADS,
AND
FLOODS.

MY WELL

most notably in the superb system which Texas has developed, are the highways planned with any thought for water, save to get it away as quickly and as thoroughly as possible. And in most places bad is made worse by the slick V-shaped ditches which are the pride of the average road gang. These cut deeper with each flood, branching back into gullies on the adjacent farms. In one county in Oklahoma there is an average of ten such gullies to every

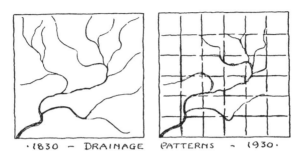

·1830 — DRAINAGE PATTERNS — 1930·

mile of rural road. Texas, on the other hand, has broad and shallow ditches, if any, and the roadsides are grassed. Surplus water is led away into convenient reservoirs, nor are the roadside ditches ever used as borrow-pits—sources of material for road building.

Beaver ponds and marshes have been drained. Thousands of miles of open ditch and tile drain have been installed to increase the amount of land for plowing. All of these devices serve to hasten water on its way, and thus deplete the reservoir beneath. The ever growing and branching systems of levees, as we have seen, prevent water from spreading out through its flood plain, and move it onward before much of it can soak in.

Finally, the fields and forests where the water falls, where its flow was filtered and ordered by a dense growth of native vegetation, and where the mellow topsoil drank it in like a sponge, these fields and forests have been cleared and put to use. Much of this use is inevitable and necessary; but much of it, too, is less valuable to society than the service once rendered in keeping us supplied with abundance of underground water, and checking the excesses wrought by high water. Needless to say, as the water has rushed away from these lands, in its new freedom, it has carried with it much of the precious topsoil, to our everlasting damage. And thus has evil grown by that upon which it feeds.

The best way to develop a proper respect for water is to live for a time where it is scarce. Major Powell, who explored the Great Plains for the government, had this experience. The officials at Washington did not. And so, when the semi-arid part of our country was divided, they ignored his plea that these divisions be made, not with respect to the compass, but entirely on the basis of available moisture.

Primitive man understood what water means. And so you will find, if you examine Doctor Shetrone's map of prehistoric sites in Ohio, that they all cluster along the once beautiful streams of that state. Moreover, the need to conserve water was one of the great civilizing influences in the infancy of the human race. The need of forethought gave birth to technical achievement. Irrigation is almost synonymous with early civilization.

Today irrigation is much in the minds of those charged with the future feeding of civilization, and justly so. But it is no panacea. Moreover, it must be invoked with caution, like the potent spirit which it is. California is happily situated for irrigation. Her soils are not overcharged with minerals, nor is the water which she uses. The supply, however, is finite, and her growing cities need it as badly as her orchards and fields; the future has to be considered.

And there is much to consider in the Great Basin and Great Plains, where nearby mountains seem to provide an endless supply of water. Here the evaporation of millenia has concentrated salts of various kinds in the upper layer of the soil, quite as it has brewed the heavy liquor of the saline lakes. Much of the mountain water is in its turn well charged with mineral salts from the rocks through which it flows. Now mineral salts in moderate amounts and suitable proportions are necessary to the growth of plants; but an excess will quickly render land unproductive. And just that has happened in more than one irrigated area, where evaporation is high and rainfall insufficient to flush the soil clean of its accumulations. Unless conditions are extremely promising, irrigation is a matter not to be lightly undertaken. And this is especially true in view of the fact that much of the land which can be irrigated is quite suitable as it is for livestock range; while new areas opened to irrigation must compete with agriculture in regions of higher rainfall.

If one were called upon to make suggestions, the most likely spot for a wider use of irrigation would be the region farther east, where rainfall is frequently sufficient for crops, but often insufficient. Across this great expanse, just west of the grain belt, and by many included in it, flow many great rivers, frequently "dry, shallow, and wide at the mouth." But often these rivers roll in great floods through the thirsty regions about them, carrying down to the Mississippi water which can be ill-spared. Some day, perhaps, means will be found to stay this water in its reckless course, and hold it for times of scanty rain. There would be little need to fear that it would injure the land on which it might be used. For that land at times receives rainfall sufficient to flush it clear of any surplus of unwholesome salts.

If such a system of supplemental irrigation could be used—not to make the desert blossom as a rose, but to stabilize organized agriculture at its precarious margin—it would be well. And if the engineers of the country could be taught to understand some of the natural processes with which they must work, in the same efficient way they are taught to calculate and construct, it would be still better. And finally, if the land upon which the water falls were so utilized and managed that the great subterranean reservoir of the nation began once more to fill up, floods would diminish in their severity and the foundations of our national life, its soil and water, would once more be secure.

Water is one of the commonest substances of our experience, yet it is also one of the most remarkable. It must be made to serve, and not to destroy us.

Chapter 6

SOIL, THE RESERVOIR OF LIFE

NOW that so many of us live upon pavement, our figures of speech are changing rapidly. We do not often hear that ancient and universal curse, that short-cut for annihilating the human without blaspheming the divine, that ultimate expression of virtuous indignation: "You are the dirt beneath my feet." The same word "dirt" is used with cheerful impartiality to indicate the soil which feeds us and to designate filth. Technically, perhaps, that is not profanity.

Soil covers the land surface nearly everywhere. Therefore it is common, vulgar stuff. We tread it down as we walk. Why give it a thought? On it falls everything dead, useless, rejected. It must be unclean. But it feeds us, and all life. In its seeming corruption is the tie that binds the endless generations into one. It surges with vitality. Against our contempt it can strike back with an invisible power more dreadful than that of ocean, for it is retributive. Yet no domestic animal responds more completely to patient, intelligent, and loving care.

Unlike the climate which goes its own way inexorably, take it or leave it, we can come to terms with the soil. For soil is above everything else a matter of collaboration. The impersonal forces of the physical world here co-labor with the organisms which people this world. Of this joint work soil is the expression. Of the energy which earth receives and life utilizes, it is the great reservoir.

Thus soil is far more than the mere litter of rock fragments and organic débris which happens to be in a particular spot. It is rock stuff disintegrated and profoundly changed by physical influences, to be sure. We shall see how completely climate writes its signature into the soil. But more than that, much more, it is stuff that has been lived in and lived upon. Animals as well as plants, the living as well as the dead—these are not simply alien objects in the soil, they are of the soil. Its qualities are an expression of their presence and their activity, no less than of the forces of climate and geological action.

And so soil is a phenomenon which represents development. The perspective of time and change are necessary, if we are to understand it. It is no goulash, to be summed up in an inventory of its contents, however useful that inventory may be. But before we look more deeply into the meaning and character of the soil, it is well to consider a question or two which the practical modern is likely to ask.

This alert individual knows that frequently the difference between a soil which produces and one which does not is merely a matter of the presence or absence of a necessary chemical substance—let us say lime, or phosphorus. Doubtless he has heard much of the promise of a new art, the so-called "tray agriculture" where glorious crops of potatoes, celery, tomatoes, can be grown in shallow pans of mineral water, and of the still more usable device of sand culture, where the same sand can be used repeatedly to produce intensive yields by simply supplying the proper nutrients. If the future will see our needs supplied through something far more efficient than natural soil, why concern ourselves with something which is, even now, becoming archaic? Why waste wealth and energy to conserve, improve, or even to understand that with which we can soon dispense?

It so happens that the men who know most about these promising devices would be the last to suppose that they will replace the husbandry of soil. Their greatest promise just now is for crops grown under glass and those which admit of intensive methods. With the sandy market garden areas of the Atlantic coastal plain now absorbing, as they often do, some thirty dollars an acre each year for fertilizer we virtually have sand culture operating out-of-doors for all it is worth. Yet the great grain farms of the mid-West are still obliged to carry on.

Supposing however that we might learn to dispense with natural soil in the production of cereals and other staples. We would still be faced with the production of fruit from orchards, timber from forests, and pasture for our livestock from the great grasslands. Space and extensive methods would still be necessary.

The earth, moreover, is our home as well as our source of raw materials. Its order must be understood and maintained. Soil plays a necessary part in conserving and distributing water, and in guarding the stability of the landscape. Our immediate surroundings cannot be agreeable, let alone beautiful, without it. Many a suburbanite struggles in vain to produce lawn and garden on the cellar diggings which his contractor has spread about a new house.

Should our hard-headed, practical questioner still remain skeptical, we might take him abroad through the countryside to see how master farmers deploy their activities over the varied soils at their disposal. He would see, not the calculated efficiency of the laboratory, to be sure, but a picture of generous and effective response to human effort. Root crops thriving in mellow sandy soil, cereals in rich black loam, orchards and pastures in clay land skilfully prepared for them—all of these represent a working relationship that can be sustained. Like the forest, they can be made to integrate with the vast symphony of nature, and must do so if man is to survive. If our skeptic sees the soil, for himself, in this light, he will see the ultimate fallacy of a human culture which ignores it. We might even clinch this point by carrying him through a land stark with the tragedy and poverty of soils destroyed.

We have said that soil must develop through the passing of time. It is not a substance, nor a mixture, but a sys-

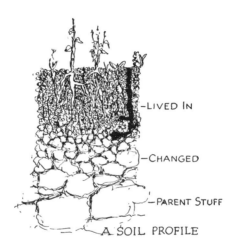

A SOIL PROFILE

tem. And it is costly stuff, too, in terms of energy. Depending upon circumstances, it is estimated that from three to ten centuries may be required for the formation of an inch of good black top soil. Such figures, however, are largely for conversational purposes. If we started from the bare rock which is the basis for soil, the time might be much longer; while peat, which has some right to be considered a soil, may form an inch in from five to twenty-five years. But no one would consider peat as by any means a mature soil, or a well-balanced one.

The real history of a soil is not to be read from its surface alone, excepting in the most general terms. We must be able to see the parent substance beneath and observe how it has been changed. Just as one slices a layer-cake vertically to discover what the baker has wrought and concealed at each level, so we rely upon vertical sections down through the soil as a means of information. These cuts may be man-made, as in road embankments, trenches, or pits; or they may be cut by nature, as they are in the exposed bank of a stream. In any event the view which results is that of the soil profile.

The profile of a soil is like a good personnel record. It reveals the parentage of the soil, and the important experiences through which it has passed in the course of its development. The art of reading this record has become highly refined, but that need not alarm us; the principles involved are reasonably simple.

The profile of a well-developed soil consists of three chief levels, or horizons. We can refer to them as the upper, middle, and lower levels. The lower level is the unchanged parent substance, such as limestone, gravel, or glacial clay. The upper level is that which has been most completely exposed to the action of climate and of living organisms. The middle level is intermediate in character. It is more weathered than the lower level, less lived in than the upper.

The transformation of parent substance into mature

soil is a long slow process, as we have said. The character of juvenile soils is determined pretty largely by the parent substance, whatever it may be. But in old soils this influence is largely offset, and the soil ultimately comes to be a rather striking expression of the climate under which it developed, regardless of the stuff from which it was first formed.

If a soil develops in a humid climate, like that of western Europe or eastern United States, it usually does so under a cover of forest. Each year the leaves pack down upon the moist surface, partially sealing it against the air, while the dense shade above keeps the moist floor cool. Under such conditions the organic material, that is, the débris of life, tends to ferment rather than to decay. The result is a more or less acid surface layer, known as leaf mold or humus.

In such a climate there is usually a surplus of rain. Readily absorbed by the leaf mold, it dissolves out some of the acid found there and soaks on down into the mineral substance below. From the upper level it removes the more soluble alkaline minerals such as lime and potash, leaving the less soluble iron and aluminum. In consequence the upper mineral level of such soils, like the leaf mold above them, tends to become somewhat acid. The stuff so removed tends to concentrate on the middle level. The animal life of such soils is largely found in the surface mold, and the flowering plants of the forest floor are as a rule shallow rooted, blooming in the early spring.

Such soils are by nature better adjusted to produce trees

than any other crop. Under cultivation the thin surface layer of organic material is rapidly destroyed, leaving exposed the upper mineral layer which as we have seen has had important nutrient substances dissolved out and carried down toward the middle level.

A considerable area of the earth's surface has had this history. Forests are frequently near the sea coast where populations are large and must be fed. Tile drains or ditches, costly fertilizer sometimes equaling the purchase price of good land not so well located, and special tillage methods must be resorted to.

Conditions are likely to be most satisfactory in broad valleys, where the normal process of soil formation has been modified by the frequent washing in of rich surface material from higher points in the watershed. At best, the farming of uplands in a forest climate must be done with skill and caution. Evidence of this is to be found in the number of farms which have been abandoned on the uplands of eastern United States, or in the forested hills of northern India and western China.

Since such soils are a major source of timber, their fate concerns others than the man who attempts to farm them. The usual routine is to clear the timber and farm the land so long as it can be made to pay even a small return. If it does not it is used for pasture; and only when the pasture becomes unprofitable is it allowed to revert to its original cover. And by that time the soil profile may have suffered to such an extent that even timber will not thrive as it should. When the soil profile is destroyed by man he is guilty of turning back the hands of nature's clock—often centuries or even millenia.

But there are serious consequences other than the reduction of timber area and the depletion of the soil when unskilful farming is attempted. The litter of the forest floor is a notable holder of water and in consequence a valuable check upon floods. Repeatedly it has been shown that the extent and violence of flood is due in no small part to destruction of this litter, aided and abetted of course, by whatever drainage may have been set under way in an anxiety to improve the moist soils for farming.

Let us pass, however, from the humid forest region to that of the subhumid prairie. Here, instead of rainfall exceeding evaporation, the two are nicely balanced, with perhaps rainfall slightly outweighed. The vegetation which expresses this condition is the prairie of tall grasses, ranging from central Nebraska eastward until it tapers out in Ohio. Southward it extends from central Texas to straggling outliers in Alabama. And on other continents it has its counterpart in steppe and grassland.

Wherever the subhumid grassland is found it becomes a region of fertile farms. To describe its rich black soil the Russians have contributed to the vocabulary of nations the word "chernozem." And from such rich black soil the nations, to a large extent, are fed. To those who control the financial destinies of civilization, events in Argentina, Australia, Russia, and our own prairie states are as important as the whims of demagogues or the progress of invention.

To begin, the roots of grasses are finely fibrous and abundant, filling the soil to whatever depth they grow. This depth is roughly proportional to the amount of rain which falls. Moreover the dead leaves of grass and other

débris do not form a wet, cold, fermenting pack like the leaves of the forest. Instead they admit sufficient air and sunlight to permit rather thorough decay, so that the humus of the prairie is well disintegrated.

The prairies are full of grazing animals which move about on the surface, enriching it with their wastes and remains, and helping to distribute materials like phosphorus, which are not always abundant in the underlying parent substance of the soil. The smaller animals of the prairie, for lack of tree homes, live largely in burrows beneath the ground—just as the first human settler is likely to make himself a cave, or at least a sodhouse. The abundance of these animal burrows, large and small, is no slight factor in moving material up and down or round about within the prairie soil. It also helps to admit air and water; and of course the débris from this animal life ultimately becomes a part of the soil humus. It is clear that the upper level of the subhumid prairie soil is a scene of intense biological activity on the part of both animals and plants. Nor should we forget that among the latter are many kinds of legumes—lupines, like the magnificent Texas Bluebonnet—vetches, clovers, mimosas, and the like, upon whose roots are colonies of bacteria that fix nitrogen from the air, thereby greatly enriching the soil.

The nice balance between rainfall and evaporation keeps the minerals of the black earth well distributed. They seldom are leached or washed downward by an excess of water as in the forest regions; nor are they often brought to the surface in great amounts as they are in dryer climates. The middle level is little more than a transition between the other two, except for the frequent presence of concretions—nodules or lumps of mineral matter, generally containing considerable lime brought down from above but remaining at hand in this second horizon.

Since the value of the prairie soil has come to be understood, it has been handled with skill and care, for there is a natural affinity between good soil and good farmers. Yet all is not well within this rich kingdom. The high cash returns for its products have in many cases led to mining and exploitation, rather than conservative care. Commercial, not subsistence farming, has been encouraged. The farmer cares more for what money he can get than for the living his farm affords. This works very well in times of good yield and good markets, but it leaves the farmer unprepared for the shock of crop failure or financial depression.

Moreover the gentle topography and the mellow, stoneless soil encourages the maximum use of power machinery, placing the soil under forced draft, so to speak. Worst feature of all, perhaps, has been the nearly complete elimination of the prairie itself—a community of plants and animals in sure adjustment to all of the vicissitudes of climate, capable at all times of building the soil while yielding wealth in the form of hay and pasture for livestock. Today the native prairie is as scarce as virgin forest, which is saying a good deal. And it has no equal as a safeguard, not only to the individual farm, but the region as a whole. Like the vanishing forest, its fate is a matter of concern to civilization.

Moving away from the region of good rainfall, the prairie passes gradually into the dry steppe or short grass, where evaporation is far in excess of the scanty rain. This in turn passes into scrub and desert, as we shall see. The corresponding change in soil profile is gradual but definite. The moist zone through which roots can penetrate becomes shallower as the vegetation tops decrease in height. The production of plant and animal material to an acre becomes steadily less, and the decay of its residues more complete, save where the great heat and dryness acts as a mummifying agent. Wind action becomes ceaseless and intense, removing fine particles wherever they are left unprotected by a cover of vegetation. The scanty rainfall is likely to come in torrents when it does occur, adding the menace of water erosion to land which is not stabilized by grass.

In arid regions there is of course little danger that nutrient minerals will be leached downward through the soil. Instead these minerals are constantly being brought to the upper level by evaporation. At its best, such soil is rich in minerals; at its worst it becomes unfit through the surplus of salts—the familiar "alkali" of the western states.

The upper level of semi-arid soils is shallow, more gray or brown than black, since there is so little residue of organic matter. The middle layer is highly impregnated with minerals, in extreme cases being almost like a slab of concrete. The "caliche" or hardpan thus formed is entirely unfit to support vegetation or agriculture if exposed by removal of the surface level above it. In North America much of the lower level is alluvial material, brought down by long ages of erosion from the Rocky Mountain area. Pebbles from New Mexico form thin layers in the hills of central Oklahoma. Such alluvial material, given adequate moisture, is of course a good parent substance for soil, as it has already been weathered and made fine.

The soil of the dry steppes is ideally suited for extensive grazing operations. Under natural conditions it maintained an extensive fauna which depended upon its sparse pasture. But it cannot profitably be handled without capital, or in small tracts; and in particular is it hazardous for agriculture. Yet its natural richness in minerals and occasionally adequate rain are a constant temptation to practice agriculture upon it. The first result of plowing is to destroy, not only the native vegetation which maintains such a precarious hold against wind and water, but the inconspicuous structure of the soil. Thereafter great skill and care must be used, not only to maintain yields, but to keep the surface level from being scalped off down to the impervious layer just below it.

At this point we may excuse the reader who is more interested in relationships than in mechanical details. With a promise to rejoin him at the beginning of the next chapter we shall take time here to consider some of the processes whose final expression is the soil profile. It should be quite clear now that not only weathered rock, but water, air, living organisms, and organic remains are all part of the soil. And while the climate can hardly be called a constituent in the same sense, yet it is quite plainly a master factor in making the soil what it is.

Climate has a great deal to do with the rate and kind of weathering of the rock substance. It also controls the water supply which is so important, and of course the abundance and kind of life which is so entwined with the history of the soil in any place. It affects the processes of decay and through them the organic content of the soil. Small wonder that the soil profile is a veritable record of climate! In Missouri and Illinois we can actually see prairie profiles being transformed to those of humid woodland, thus confirming the evidence from other sources of a slowly changing climate.

By the process of physical weathering, rock is crumbled into fragments known as gravel, sand, or silt according to their fineness—the silt being virtually a powder. The size of these particles has a great deal to do with the rate of water movement and the amount of water held. Sand into which water moves rapidly will contain less water when saturated than will silt, and silt gives up water less generously to plants. In consequence plants will wilt sooner in silt soils than in sandy during the heat of summer. Likewise the size of soil particles affects the amount of air which can enter. Root crops which require much oxygen for their rapid underground growth will thus thrive best in sandy soils.

But rock materials may undergo further change than mere physical crumbling. So-called chemical weathering either converts them into a paste or glue, known as clay, or into dissolved material which becomes a part of the soil solution.

The clay particles are too fine to be seen with a microscope by ordinary methods, but too coarse to be in solution. Clays swell when wet, are sticky, and cake when dry. At all times they hold water tenaciously—even more so than does silt. They are often abundant in the middle level, making that layer hard to handle when it is exposed. In the upper level, if not too abundant, they serve the beneficent purpose of holding the coarser particles and the organic material in crumbs or floccules, with great advantage to the soil texture. Other things being equal clay formation is in proportion to the wetness and warmth of a soil. It may reach an undesirable extreme in the tropics. Moreover in the tropics the great heat causes organic material to decompose so rapidly that there is not enough humus to offset the bad effects of too much clay. Thus we have the spectacle of abandoned land and technical troubles in the banana plantations of today no less than in tropical civilizations of the past.

In temperate regions the clay content of a soil may increase with its age, but so does the amount of humus, and thus the difficulty is avoided. For while humus, like clay, represents the colloidal phase of matter, it is mellow instead of caked when dry. It holds much water, but it likewise admits air readily, and it also furnishes the medium in which beneficial bacterial activity can proceed.

We have said that a second effect of chemical weathering is to produce the dissolved minerals of the soil. For a long time these were thought to be the main secret of fertility, and with water to be the entire basis for sustaining life. Today we know that a third factor, that of soil structure is quite as important, for to a large degree it determines the availability of minerals, water, and air, as well as a number of invisible processes which play a part in supporting vegetation and animals.

Yet the soil minerals remain extremely important. Less than a score appear to be essential to living things, not more than about six create practical difficulties by their absence, and three require critical attention at all times. These three, nitrogen, phosphorus, and potassium, whose symbols appear on every fertilizer sack, loom large in the affairs of nations, no less than in the troubled routine of the individual farmer.

All are present in the soil in the form of very soluble compounds, easily washed out and lost. All are removed in large quantities when agricultural products leave the farm. All are expensive to obtain commercially: $30 an acre per annum goes into the purchase of them in market garden areas of the eastern coast. All are important in chemical industry, particularly so in time of war. The same plants and materials may be used to produce dyestuffs, fertilizers and explosives. War is not only the great destroyer of what the soil produces, but the insatiable rival of the soil for needed chemical materials. Civilization cannot strike a balance with nature until it comes to realize this fact.

Fortunately the air is a great reservoir of nitrogen, although the gas which composes four-fifths of the atmosphere is not available directly to plants and animals. It must first be fixed by bacteria—a process which occurs in nature or under good land management, but not otherwise. Ordinary exploitation of the soil strikes directly at this essential and orderly change. To some extent—and expensively—the mischief can be offset by electrical fixation of nitrogen, now an established industrial procedure. Once in the soil, nitrogen compounds, like those of other minerals, may be husbanded and used repeatedly by successive generations of plants and animals; or of course they may be, and often are, wasted and lost.

Phosphorus and potassium are to be obtained from mineral deposits. More than once, as in Strassfurt and Chile, such deposits have been the subject of acrimonious national rivalry. At best they are not too cheap or abundant. When necessity arises, the sea may be drawn upon to secure them. It is of course the great final repository of minerals washed out from the land. Fish are rich in phosphorus, seaweeds in potash. But neither is a cheap source, nor so available as plant and animal residues produced on land. At present much of this latter material is blithely poured into drainage and away, ultimately to enrich the sea, of course, but meanwhile to poison the streams and

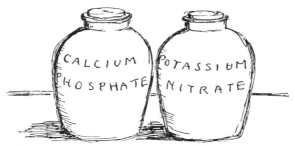

PEACE OR WAR — THESE ARE THE INGREDIENTS

impoverish the land. We came from the sea. Unless we watch ourselves we might have to return to it on the heels of our hard-won soluble mineral wealth!

There is wisdom, crude though it be, in the story of the old continental farmer who drove his marriageable son past a long row of peasant homes. Finally he paused before one, saying "Da ist der Mist, da verheiratest du dich." ("There is *the* manure pile—go there for your wife.")

We have referred in the preceding pages to the living organisms which are in the soil. It should be clear by now that bacteria and roots, worms and rodents, are as much a part of the soil as are the minerals, the humus, the water, and the soil atmosphere. The thoroughness with which the soil is permeated by roots cannot be believed by one who has not seen root systems exposed for study by painful picking or by washing. Even more intensively is the soil pervaded by microscopic life, both plant and animal, while larger animals such as worms, insects, and burrowing vertebrates are quite as much a part of it.

Any city is more than the houses, the streets, the utilities, the people, or even the institutions which compose it. It is the integration of all of these factors. Just so is the soil—the integration of all of the physical and organic factors which make it what it is. As the city grows and changes, assuming the intangible thing called character, so the soil develops and assumes character until it is in a situation approaching dynamic equilibrium of all of the factors which influence it. It comes to express the entire complex which impinges upon it. Small wonder that in this Dark Mantle there lies so much, not only of the past, but of the future course of this earth and those who dwell upon it!

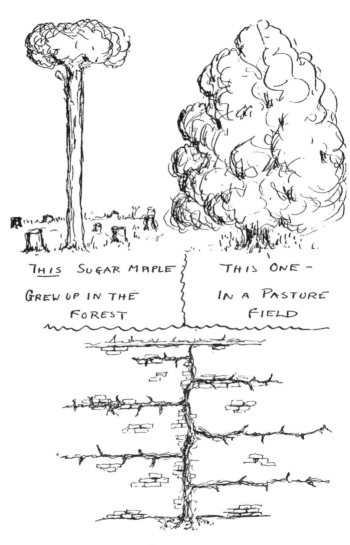

THE PATTERN
OF LIVING NATURE

Chapter 7

WHAT IS LIFE?

WITHOUT the proddings of priest and philosopher, most of us are too busy trying to hold on to life, enjoying it as we can, to wonder much about its meaning. Once we win a little leisure and security, however, we require a most elaborate and artificial routine to keep our minds off the haunting and troublesome question of life and its significance.

The facets of approach are as varied as experience itself. Anything which can possibly happen to us may illuminate the problem. But with the coming of the New Age, we hearken more and more to the man of science, forgetting perhaps that he too may approach the problem in manifold ways.

Some years ago, Dr. George Crile of Cleveland, as

skilled in stirring up excitement as he was in mending bodies, allowed the public to look through his microscopes. Therein could be seen tiny masses of material moving about, growing in size, and at times dividing themselves into smaller pieces. To the unpracticed eye it was all a very good imitation of life. Furthermore it was the result of putting together certain chemical substances which were not alive. This was too much for the newspapers to pass up, and they did it ample justice.

On the train leaving Cleveland after this exhibition, I was approached by a fine looking man who had chanced to learn that I was a student of life. He told me that he and his wife very much wanted to talk to me, so I joined them. They wanted to know if Dr. Crile had really created life by artificial means. They had just lost an only son, nearly grown. Were he and all that live really as precious and unique as they imagined, or did he merely have the significance of another piece of sheet metal, turned out in endless and meaningless duplication by the stamp mill?

So far as any artificially-made life was concerned, I was able to answer their questions. The stuff under the microscope had been soap in the process of formation, showing what are known technically as myelin-forms. The appearance was lively enough, and in some ways different from any myelin-forms I had ever seen; but that was due to the peculiar combination of fats and alkalis employed. And the only thing alive about the experiment was the assistant who set up a fresh mixture when the old ran down. The phenomenon showed just what its exhibitor claimed for it, and nothing more—it exemplified some of the physical properties of living cells. And of course it left the eternal problem of human values and responsibilities just about where Moses and Buddha and Socrates found it.

The system lacked continuity, among other things. It could not maintain itself in a balanced relation to the outer world. It could not build experience into itself, except in a very feeble sense. It was a manifestation of the life of a very clever and capable man, but not of its own life.

It lacked precisely those attributes which make life a significant thing—continuity, experience, adjustment—in relation to the world about, and running through all these the thread of slow, inevitable development and change. Such is the framework in which life expresses itself. Neither to individual nor his group is there any promise of permanence—only the chance of achieving relative order and stability. And to mankind, with all of his values caught in a social order, this chance is the hope on which everything earthly hangs; so it behooves him to know the terms he must meet. As we go in search of illustration we can discard the microscope and the shelter of the laboratory.

Here and there, at long intervals throughout the eastern states, are tiny fragments of the virgin forest, sheltered and preserved by some miracle either of diligence or its contrasting vice. Compared with the fields about it, such a forest is like the populous hearthside of an ancient peasant cottage; the fields too often suggest the man-woman-poodle futility of the modern apartment.

Step into one of the islands of the primeval. You enter

another world. The soft, rich, resilient carpet of dead leaves gives beneath your feet. The sun which drives us into smoked glasses whenever we go abroad, is absorbed by the heavy green canopy above you, and its energy put to work. The light which reaches through is soft, the air cool and humid. You are in the midst of a wilderness of trees, bewildering, chaotic.

But do not be deceived. You would not judge the order of the peasant cottage by the tumbling, shouting children. Go to the elders. Search out the oldest, largest, trees of the forest—white oak, sugar maple, beech. You will find them spaced apart, standing with rugged dignity, sharing their domain. These biggest trees give you the true perspective of the forest. About the base of each is a charmed circle in which only the humblest mosses and herbs are found; if any seedlings of trees appear there, they remain dwarfed and depauperate, as though overwhelmed by their nearness to majesty.

But in the neutral ground where light filters between the great crowns, the seedlings and saplings are crowded, racing their growth towards the top. And the ripening of an elder giant is an event which makes history in the forest world. Slowly the processes of destruction and decay overtake that of growth. Insects and fungi break down the stout old heart, branches die and admit the sunlight, calling the suppressed growth below into new activity.

In the course of time, the dying giant falls. Its prostrate trunk molders away, first covered with bracket fungi whose beautiful variety of soft colors are the despair of dyer and painter. Later the log turns a rich red brown, giving foothold to mosses, ferns, and the delicate herbs of the forest floor. Gradually its substance merges into that of the soil, marked only by a long low mound of lusher growth than that around it. There is life abundant, even in death, within the forest.

We are too late today to witness the animal life which once fitted into this scene. True, the insects, birds, and smaller animals remain, but the larger are mostly gone. Each has, or had, his own niche, from the fungus gnats depositing their eggs in each mushroom as it grew to maturity and served its function, to the deer browsing upon the shrubbery and lower branches of the trees. Less than a lifetime ago it was possible for the farmer to ask his wife how many squirrels she needed for dinner and meet her wishes exactly; passenger pigeons, feeding in flocks upon the beechnuts, could be killed with clubs or rocks. Wild turkey abounded, in addition to the mink, groundhogs, and ovenbirds which remain.

Through uncounted centuries this teeming, balanced system was maintained. At times its boundaries shifted with changes in climate. Fires or the casual interference of the aborigine upset its order here and there, but not for long. Then we came, with our livestock, our rifles, our axes and our plows. We created conditions under which the primeval forest could not maintain its balance.

As we cut the older trees, we stimulated the undergrowth by admitting light. It was no longer possible to see a deer at five hundred yards beneath the canopy; so we burned out the brush, thus killing the seedling trees and destroying the leaf mold. Weeds of the field, vulgar noisome importations from the crowded Old World, replaced the delicate orchids, maidenhair ferns, wake-robins, and spicebush. Or if we turned our cattle under the trees in place of the exterminated deer, they cleared out the forest growth. The less desirable old trees might hang on for a time, but the death sentence of the forest had been pronounced.

And so, about the margins of its rare remnants today, we step into fields of bluegrass and clover, spotted with Canada thistle, vervain, ragweed, mullein, and the like. Grasshoppers, fieldmice, and meadow larks rule in place of the stealthy forest animals. Even if the field has been skilfully tilled, the old mat of leaf mold is gone, and a new system of life set up—one which requires constant thought and tending to produce a tithe of the life which the forest maintained as a matter of its own routine.

We are not concerned with which is the better. Much of the forest had to go in order that we might live. But let us remember, as we drive from St. Louis to Baltimore, that a world of living, and seemingly permanent, reality has vanished from the landscape. A new world of our own creation has replaced it. What is our guarantee that it can last? Like the world of the forest, it cannot long exist in unbalance.

Historical romance is a questionable means of learning the truth of the past. It is as great an indignity to the dead as to the living when we read their thoughts and motives to suit ourselves. Yet such fiction serves at least one useful purpose, aside from entertainment. Steep ourselves in it, and we come gradually to realize that each human culture looks upon itself as a world of permanence and stability, except perhaps in some period of supreme catastrophe like Southern Reconstruction or the weary retreat of the Red Man.

With our imagination in good vigor, it is even better to take our history straight from its sources. Any good museum abounds in pieces of exquisite, unhurried craftsmanship, evidences of deliberate living and ritual from ancient and vanished sources. It makes little difference whether we are interested in carved gems or burials, hieroglyphs or tobacco pipes. All tell the same story. The maker knew that he himself would pass on, but it was inconceivable to him that the culture of which he was a part would disappear.

Who among us would have the courage to found great endowments, design magnificent buildings, or strive for perfection in anything if he knew that the next four hundred years would bring us what the past four hundred have brought to the Aztec and Inca?

One of the interesting and significant documents of antiquity is the Code of Hammurabi—the legal system of Mesopotamia as it was drawn up just about four thousand years ago. It was, of course, the expression of a slow outgrowth from more primitive legal systems among the very ancient Sumerians, quite as later systems have borrowed heavily from it.

This great Code contains the most complete and detailed regulations for the governance of human relations in a busy country, dependent upon irrigation and commerce. Property and other human rights, maritime law—even price-fixing—are dealt with almost minutely. One feels in reading it that its authors were living in a quite substantial world, where tomorrows in endless succession gave promise of being as today. Society was ordered, each man knew his place as his fathers had known theirs, and sons to come might expect to follow on.

Of course the Mongol rulers of Persia destroyed the great irrigation works which kept Mesopotamia from being a desert. This happened in the thirteenth century and today the Near-Eastern governments are only too glad to exchange oil for water expertise. The Persian ruthlessness wiped out the physical framework to which a long and powerful culture had fitted itself. Wandering, bickering herdsmen took over the buried ruins of Kish and Babylon and Ninevah.

Repeatedly before that, however, the smooth order which Hammurabi knew had been shaken to its foundations. A life-giving river might shift its course, leaving some great city without water, as Kish had been left. Salt might accumulate until irrigated fields would no longer produce. Silts and muds washed down by the streams filled great navigable lakes and pushed the sea away, changing the pattern and means of commerce. Strong cultures grew up nearby, with superior technical means at their disposal, giving them an advantage in peace and war.

Whenever any of these events took place, the old smooth order became disorder, and harsh adjustment was necessary. But until the great destruction, some adjustment was always possible. The entry of the Khan was the death sentence, just as the entry of the white man had doomed the hardwood forest in North America. Life lives in a changing world and must be ready for change—orderly change —if it is to carry on. The man who calls treaties "scraps of paper" is philosopher, whether or not he is a scoundrel in the bargain.

Thoughtfully, some one had inscribed the Code of Hammurabi, not upon paper or parchment, but upon a great table of granite-like rock which was discovered at Susa early in the present century.

If the perspective of human history is not convincing, one has only to go back into geological time, as we shall do in the next chapter. Read of the rich flowering of life in the Devonian, the Coal Age, and still later in the Cretaceous. Ask some geologist friend to show you the casts and fossils of once living plants and animals which have come down to us from each of those three periods. Ask him to show you some of the notable paintings or drawings in which landscapes from these periods have been reconstructed with meticulous care, with the weird and bizarre denizens in place. You will see trees, shrubs, beasts which were once as substantial as those which surround us. Doubtless your first thought will be: "These had to pass, because they were not so perfect as those of today." In a measure that is true, but it is only part of the truth.

Quiz your geologist friend still further, and you will find that the earth and its climate underwent profound changes to which the older forms of life in their turns were unfitted. And so, unless they remained flexible, with a capacity for change and adjustment, they were wiped out—became extinct.

Whether our perspective is on the grand or trivial scale, we find life to be balance. The tree-fern cannot survive an era of cold and deserts; the dinosaur is doomed in a world of quicker, more intelligent, if weaker beings; the flint worker recedes before the coming of sharp steel. To study the character and mechanism of this balance is our task in the succeeding chapters.

And lest the prospect seem utterly dismal and cheerless as we move along, let us remember that man has means at his disposal which go beyond the cleverest works of his hands. These means he has scarcely tapped. In his use of them his destiny lies.

Chapter 8

THE PAGEANT OF LIFE

THE puzzle of the earth's history reminds one of those cleverly built Oriental boxes which open with such surprising ease when the secret spring is touched. Land-forms and their constituent rocks, seemingly a scrambled confusion, became orderly and legible by the magic of a very simple idea—that of uniformity. As soon as the geologist began to assume that mesas and valleys, sandstones and shales, had been formed in the past by the same processes that now operate to produce them, the very earth about him became luminous and vocal with its story. And today the pious literalist who would have all of us believe that Genesis is a notebook instead of a magnificent poem must concede that geology is at least a workable subject; the oil wells which build his churches and pay his taxes are stubborn realities. Even the boys on the street know how such wells are located, or at any rate have a wholesome respect for the craft of the "rock-hound."

Yet this uniformitarianism, which started as an assumption and has earned its right to be called a great natural law, does not imply by any means that the earth and its inhabitants have always been what they are today. The operation of forces, and the behavior of matter in nature may be changeless, but there is no sameness or monotony in their visible expression. Each morning's sun, almost literally, looks upon a new earth.

It did not take the geologist long to realize that the pattern of land and sea had gone through an amazing series of transformations. Old ocean bottom can be found in many places some thousands of feet above the present sea

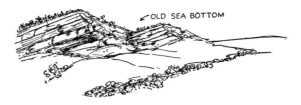

level, while the substantial real estate of other days is now the playground for fishes. The geologist knows better than anyone how passing a thing is the immediate landscape.

Moreover his associate, the student of fossil plants and

animals, reminds us that the living part of the environment has moved along with the course of time. It is a very different thing to be alive in a world of trees, grasses, and warm-blooded mammals than it would be to find ourselves in the midst of great forests of ferns, peopled and ruled by reptiles. Life has changed with the changing of the landscape and the shifting of the continents.

So much for landscape and life. Climate remains; what of it? So far as any particular spot is concerned, the rocks themselves bespeak constant change of climate. Today a dry, sparse, and scrubby forest may cover shallow seams of coal which could only have been formed in some dark and ancient swamp in a region of high rainfall. Or oyster shells may crop to the surface in the Wyoming hills and form little spots of desiccated alkaline desert where once the air was muggy above the brackish covers of ocean. Certainly the climate of any particular spot must change with its distance from the sea and its position on a changing continent.

This is not, however, the whole story. The red-beds which stretch like a pink girdle across our southern states carry the record of great continental deserts. At other times coal was formed over large areas of the earth by the piling up of luxuriant vegetation that was the antithesis of desert conditions. And masses of earth and rock, hundreds of miles from their point of origin, once piously called "diluvium" and attributed to the Noachian disaster, are now known to have been scraped along by continental icecaps of inconceivable bigness as compared with the shrunken polar snow fields of today. There is plenty of reason to know, in other words, that at various times the climate of our planet as a whole was shifted in one direction or another. And even during periods of relative stability, such as the past ten thousand years, we have discovered that climate has gone through marked changes, less revolutionary of course, but sufficient to make a difference in the boundaries of forests and plains, and in the fortunes of bison and men, horses, camels, and pine trees.

Science will not rest, of course, while these intriguing problems of the past afford a single clue whose meaning remains unsolved. This putting together a story after it has happened is really glorified detective work, hazardous and delicate, carried on by indirection and inference. It is not a task for those who seek pontifical authority or who dread the havoc which new evidence can wreak upon beloved theories.

Nor, let us remember, is it a mere selfish indulgence of idle curiosity. Society has a stake in the search. Concerning the present there is much which we shall not know without a more complete knowledge of the past. The disastrous drought of the 1930's reminds us of the need for a better understanding of the rhythm of climate. The time-table of the earth and its changes may not interest everybody, but it certainly concerns us all. The problem of human culture, its origins, movements, and characteristics is coming increasingly to be a problem of environmental change. There is an archeology of environment no less than of flints, pottery, and bones.

The drainage ditches and dumps from a modern city may be unsavory places, but if the city were to disappear, they might constitute its only record, and a useful one it would be to the trained eye. Similarly the beds of sediment washed from ancient continents are today our chief source of information about them. The character and disposition of these sediments tell us much about the former land surface and the conditions under which its population lived. Slow sedate streams flowing through broad valleys carpeted with rich soil and abundant life bring to the ocean very different materials from those scoured down by the infrequent, ungovernable washouts of a dry, inhospitable land. But mainly we depend for our knowledge of past conditions upon the remnants of life itself. We can reconstruct the course of life from the fossil remains of such organisms as have been suitably preserved.

Even with a fairly satisfactory knowledge of fossil remains, there are of course dangers in deducing too much about the conditions under which their original possessors lived. Who, knowing where the elephant thrives today, would guess that exploration would yield the frozen carcasses of his close kin in the barren arctic wastes of Siberia? And even the professional botanist gets a jolt from Professor Seward's photograph of tree ferns which are growing today where a glacier is in sight in the southern hemisphere. Yet interrupted, dim, and deceptive as our glimpses of past life may be, they tell us much. Very briefly, then, let us examine some of the higher spots in the record.

Before we can discuss a few of the great problems of adjustment between life and its environment which have been encountered along the way, it is well to see something of the order in which living things developed, and the kinds which dominated the earth at different stages in the past. Hit-and-run tourists that we are, we shall have to use a guide. Here it is, the time-table used by geologists, in a very much simplified form. And if, in the succeeding pages of this chapter, we become lost, we can turn back to it.

A SIMPLE TABLE OF GEOLOGICAL TIME

Here the whole of geological time is divided into five great eras; these in turn are divided into periods. Memoranda of the fossil record are added here and there as guides.

Era   Period

Cenozoic—Quaternary—Recent—Modern Man
      (30,000 years)
   Pleistocene or Glacial—Early Man

 Tertiary—Pliocene
    Miocene—Three-toed horse—pre-human anthropoid
    Oligocene—Specialized mammals
    Eocene—Primitive mammals, vegetarian and carnivorous

Mesozoic—Cretaceous—End of the great reptiles—modern vegetation—primitive birds and small early mammals
　Jurassic—Reptiles rule the earth
　Triassic—Origin of the flowering plants

Paleozoic—Permian—Conifers replace the ferns—primitive reptiles
　Carboniferous—Ferns clothe the earth—insects, amphibians, very early reptiles
　Devonian—Great forests of ferns and primitive conifers — first land vertebrates — rich invertebrate fauna—many fish
　Silurian—Early simple land-plants—primitive fish
　Ordovician—Great reefs of simple seaweeds
　Cambrian—Invertebrates only—marine forms (Perhaps a billion years ago)

Proterozoic
Archaeozoic } Pre-Cambrian—Simple marine plants and invertebrates—scanty remains

This table tells us some interesting and significant things at a glance. We see that human history is a trifling fragment of geological time. Life of some sort has been present upon the earth for a period of almost inconceivable length. Speaking bluntly, it has existed long enough for almost anything to have happened.

And we can see that much has happened. The actual beginnings of life itself are lost; but there is plenty of evidence of a steady advance from the greatest simplicity to an amazing complexity. The earliest remains are those of

PLANTS BEFORE ANIMALS!

soft-bodied plants and animals. Firm skeletons and the organization thereby made possible were a matter of gradual achievement, of repeated, not always successful, trial.

The record is full of great groups that rose to power and abundance, only to recede and disappear in the face of their own followers, like successive chords in some titanic modulation.

The table tells us also of the earlier enrichment of life within the sea and its prolonged struggle thereafter to move upon the land and become adjusted to all manner of conditions there. And since the sea is far more constant than the land, the striking changes of the latter part of geological time largely concern the life on land.

We also see in this simple outline of events that animals in the past, as now, have been dependent upon plants for subsistence. Each great wave of animal development is preceded by an appearance of the appropriate and necessary plant life. Land flora came before land fauna. The flowering plants arose ahead of the mammals. And it is pretty clear from the table, too, that the passing of time has tightened the fabric of interdependence among living things. The half-million kinds of insects now in existence are involved with each other and the rest of the living world beyond any power of separation or retreat.

As for man, most recent and most specialized of all organisms, and in his own conceit most powerful, he is the spoiled darling of the whole system. Because he can utilize and apparently control all other organisms, he is likely to forget that he is the beneficiary of a prolonged and delicately balanced development. Thus far in his history he has done more to destroy than to perpetuate that balance. And the deep undertones of the scientific record reënforce the somber words of prophecy, "Let him that standeth take heed lest he fall."

With these very general facts about the past of life, let us retrace our steps examining certain phases of the scene in greater detail. As we do so, let us bear in mind that the record of the fossils is sketchy and imperfect. What proportion of our worthy citizens have access to the august portraits of their eight great-grand-parents; indeed, how many know the names of these dignitaries? An expert on heraldry states that not more than a handful of families in England can trace an authentic lineage a thousand years, and even among the lines of these exalted few there is no doubt plenty of shadow and mystery. The ancestral record of living organisms goes back some hundreds of millions of years. Only by the luckiest coincidence is an individual preserved and later revealed to the eyes of science. The remarkable thing is not that there are vast gaps in the story, but that it hangs together as well as it does.

So far as beginnings are concerned the record is tantalizing. How did life first begin? What was the first vertebrate, the first flowering plant? There is little comfort for those who seek an answer to such questions. We seldom find fossils of anything but groups which have been long and well established and whose individuals died in great numbers under circumstances favoring their preservation.

Calendars however, whether scientific or ecclesiastical, must have a point of beginning, and so the calendar of fossil time begins its great primary era, the Paleozoic, with the Cambrian. But it must not be thought that the Cambrian is therefore the time when life began. Far from it. The Cambrian is merely the period whose rocks are marked by the first abundance of fossils. And while these fossils consist of invertebrate animals and seaweeds, they are all well enough developed and far from being really primitive. Some Cambrian trilobites might deceive a nearsighted gourmet into thinking them shrimp or crayfish. Marine plants abounded, and the seas of Cambrian time were swarming with animals having special organs of locomotion, digestion, sense-perception, and nervous systems to coördinate them all.

Looking backward from this zero hour we find only scattering fossils. Some structures indeed resemble the tufted or massive seaweeds of today, but we are not too sure that these are really fossils and not ancient fairy gardens like those Jack Frost makes with ice crystals. If simplicity is any guide to the earliest life, we can draw up specifications; we know the viruses today which have no visible organized form but which can increase their own substance, that substance being protein-like. We also know bacteria which can manufacture their own food without the aid of sunlight or the green pigment in leaves. That is as far as we can go. We do not even know whether such archaic life came into being in shallow seas, or in nutrient-rich pools, somewhere on land. But living substance today requires the salts found in sea water, and certainly the sea has been, if not the womb, at least the cradle of early life.

Before the middle of the primary era, a balance was established within the ocean which has persisted without great disturbance until the present. The hungry life of ocean is fed by sunlight. Since this light diminishes rapidly with depth, the surface is a welter of floating life known as plankton. Myriads of microscopic plants convert minerals, water and carbon dioxide into food, upon which countless invertebrates feed. Upon these in turn the larger forms, predominantly the vertebrate fish, are fed. This plankton is the great pasture of the sea, and its débris, constantly settling to the bottom, supports animals at greater depths, meanwhile helping to supply the sediments which in time become rock.

If the plankton is like pasture, there is anchored life in shallow waters which might be likened to forests. Great reefs of lime-secreting algae and other seaweeds have, since early geological time, afforded shelter and food to animal life.

Since its establishment this pattern has become enriched and altered in detail, of course. Primitive fishes have been replaced by more specialized types, while mammals and flowering plants originating on the land have sent representatives such as the whale, the seal, and the ribbon grass back into the ocean. Trilobites have been replaced by modern crustaceans. The life of ocean today has flowered into an amazing richness of form, yet on the whole remains at a restrained level of organization. Both animals and plants show a great development of microscopic, small and moderate sized organisms, highly interdependent and affording means of support to those most primitive, as well as most ancient, of known vertebrates—the fish.

Endless modifications of structure and function are involved in the ancient and intricate pattern of ocean life. Bottom feeders tend to be horizontally flattened, sluggish, often armored on the back. Rapid swimmers, notably the fish, are beautifully and effectively streamlined. "Tail like a mackerel, head like a cod," is still a boat builder's formula. Sedentary forms, such as molluscs, sponges and corals, are actively moving during their juvenile period and thus migrate and spread. Floating plants growing submerged may have special pigments which help absorb the scanty sunlight that filters down to them.

But the great primary era, or Paleozoic, which witnessed the conquest of the ocean by life, also saw another great adventure set going. This was the invasion of the land, where conditions are not as constant as within the sea, and where hazard is present at every step. Early Paleozoic fossils of the Ordovician and particularly the Silurian include bizarre plants whose fitness for land becomes more pronounced with the passing of time. Certainly these were not far removed from the soft-bodied seaweeds. But behind a veil we have not yet penetrated there were developing races of mighty, if primitive, trees. Without known precursors we see these forests burgeon forth in the mid-Paleozoic or Devonian where their trunks, sometimes five feet in diameter, are abundantly preserved. And peopling these forests were the first insects, land molluscs, and those first land vertebrates, the amphibians. The problem of a bulky, permanent land vegetation was on the road towards a solution. But as yet there was no evidence of an animal population commensurate with the organic material being produced.

As the Paleozoic moved past its zenith of the Devonian, there occurred the dazzling afterglow of the coal measures, or Carboniferous time. We have little means of knowing what was taking place on the high ground of the continental interiors, but in coastal swamps and on low ground there were amazing forests of ferns and other weird trees without true flowers. Accumulating more rapidly than they could be decayed, their remains furnish many of our present coal beds. Among dank recesses the insects and amphibians continued to thrive, while from the latter a higher form of vertebrate animal, the primitive reptiles emerged. These reptiles laid their eggs upon land instead of in the water, and were land animals from the moment of hatching instead of spending a period of juvenile probation in the water as did the amphibia. The emergence of the higher animals was now complete. Henceforth individual species might return to the water to live, but the group as a whole had been graduated onto land.

When the Carboniferous period was succeeded by the time of cold and deserts—the Permian, which ended the Paleozoic—it became evident that the mighty ferns and their kin had not solved the problem of surviving the hardships of living upon land. Their great, delicate-tipped fronds, their unprotected growing points, their complicated, uncertain method of reproduction, and their limited facilities for thickening and branching became a liability. They disappeared, practically speaking. But alongside of them the ancestors of the modern "evergreens" and other cone-bearing seed plants weathered through, for they were better protected at every stage of life history.

Thus the secondary or Mesozoic era was ushered into a world landscaped, so far as we know, by cone-bearing trees and peopled by the vigorous, youthful group of reptiles. Gradually during this era the climate of the earth again became genial as it had been in the Devonian and Carboniferous, so that the Mesozoic era ended with another luxuriant outburst of life—the Cretaceous.

Meanwhile the reptiles had multiplied until they possessed the earth and to an extent its waters and the air above it. Powerful, sometimes swift, and in many ways efficient, they nevertheless bore the seed of their own destruction. Their body temperature was not effectively regulated, so that cold rendered them sluggish. They generally laid their eggs in the ground and gave the young rather scanty attention. Equally important, their central nervous system was poorly developed, inadequate to control the vast mechanical system of their bodies.

Throughout the Mesozoic, the plant world had become enriched by the development of flowering plants, including many kinds which still persist about as they were then —the oaks, hickories, magnolias, sassafras, to name but a few. And these flowering plants were perfectly suited to nourish and shelter, not only the growing hordes of insects but the warm-blooded birds and primitive mammals which put in their appearance at the end of the Mesozoic, that is, in Cretaceous time.

These small pioneer mammals were literally in at the death of the great reptile groups. It is often suggested that along with their greater intelligence and more continuous activity, they may have possessed a penchant for sucking reptile eggs, as well as an ability to hide out from the ubiquitous monsters who laid them. Thus plausibly may we think of David and Goliath or Gulliver and the Lilliputians foreshadowed in the unsmiling humor of the ancient cosmos.

At any rate the Tertiary division of geologic time began in a world of flowering plants and primitive mammals, with the earth cleared of its saurian dragons. And as time proceeded this period witnessed the prolonged and classic history of the horse and the more obscure birth of other modern mammals. Undoubtedly the development of great continental grasslands or steppes was a potent factor in these events, but unhappily for us much of that side of the story is lost. We do not know when or how the grasses, so admirably suited to dry climates and to grazing, had their origins. We have to work by inference, based upon the teeth and other anatomical features of such fossils as we have found. Among other animals of the late Tertiary was the precursor of man.

As the Tertiary was finally replaced by the Quaternary in which we of the present live, the earth had approximately its present geographical pattern, substantially, although by no means all of its present vegetation and animal life. For the Quaternary has been a period of great climatic stress. Not once, but repeatedly have the great polar ice caps so thickened that their enormous weight caused them to spread in a grinding, viscous flow until they reached deep into what are at present temperate climates. Seemingly enough water was impounded in the frozen masses each time to lower the sea level considerably. And each advance of the ice caused, we know, profound shifting of the plant and animal life in its pathway. In Europe this resulted in impoverishing the rich Tertiary life, for mountains cut off retreat to the south; but, in North America the avenue of escape toward the south was open and with each recession of the ice, the displaced communities came back close behind its melting edge.

Between advances have been interglacial epochs resembling the present, and of which the present may well be but one. Obviously such profound oscillations of climate have played a large part in producing the present pattern and composition of life upon the earth. Surface changes produced by the ice often created barriers, isolating groups and favoring minor evolutionary developments. And the youthful topography, usually rich in minerals, provided a variety of habitats suitable to a wide range of organisms.

Man was on the scene not later than the third interglacial epoch. Aided by his superior intelligence, such as it is, and by the marvelous inventive genius which his hand has made possible, he survived—to begin his slow battle for dominion of the earth in postglacial time. Fitting at first into living communities of plants and animals without essentially altering their balance, he has emerged into dominance. So complete is his control that in most parts of the earth today he is surrounded by communities which have been profoundly modified by his own interference. For the first time in its geological history, earth is overrun and ruled, not by a group of organisms, but by a single species. Will this species avoid the fate of the mighty groups which preceded it?

Chapter 9

PLANTS, LIFE AT ANCHOR

AS YOU read this title, the odds are that you will think of green leaves, stems, flowers, and roots. The most familiar plants which are everywhere about us, and which furnish the usual background of our thoughts are green and leafy. They possess stems to support their foliage in the sunlight and to connect them with the hidden roots. Their seeds are enveloped in a fruit, product of their flowers.

To such plants, mainly, man like other animals goes in search of food, of drugs, and of structural material. Upon them even the flesh eaters depend, for if we follow the chain of food to its final source, that source is ultimately the food manufactured by green plants.

> "A bee flew down and ate an ant,
> A bug he ate the bee;

A hen then gobbled down the bug
  But failed the hawk to see.
The hawk had eaten up the hen
  Before he saw the cat
Which ate him up, but then a dog
  Ate pussy quick as scat!
A wolf now sprang upon the dog
  And ate him in a trice,
And then a lion ate the wolf
  And found him very nice.
But when the lion fell asleep
  He said, 'I really can't
Imagine why that wolf should taste
  Exactly like an ant!' "[1]

Had the author of these immortal lines started with the fact that the ant was a vegetarian, his case would have been invulnerable!

What is a plant? A cow is an animal, and an apple tree as certainly is a plant. But why? If we pass to less familiar forms of life, such as the coral and sponge, perhaps even the toadstool, sharp eyes and good judgment are required to put them in the proper company. Should we descend the ladder of life still further into that nether world of microscopic beings, we are obliged to call upon a solemn jury of technical experts to say that this is a plant, that an animal. And—breathe it not beyond the walls—

[1] From L. Frank Baum, *Father Goose His Book*. Pictures by W. W. Denslow. By kind permission of the publishers, The Bobbs-Merrill Company, Indianapolis, Indiana.

there are cases like that of the slime-molds and the tiny, lively green Euglena in which the jury itself has been hung!

Within parenthesis may we add that in such situations lies one of the great stumbling blocks between the man of science and the man of affairs. The latter, particularly where he works with public opinion in the mass, must have, not finely tempered tools, but bludgeons. One crude slogan is worth a dozen delicately balanced ideas.

Resuming work on our puzzle, it begins to resolve somewhat when we recall that plants and animals probably had a common beginning and have been growing, each in their own way, ever since that origin. Our own kind of animal, with his names for everything, has been one of the last to appear upon the scene. Seeing about him these two highly developed and clearly different kingdoms of life, he coins a word for each. Thus the idea of a plant is far older and more practical a matter than any precise understanding of what a plant is.

Although most familiar plants are green, many are not. The waxy white Indian Pipe, the golden dodder, the pallid, deadly Amanita mushroom, and the grayish mildew all are plants, but they do not make their own food. Only plants which are green can do so. Nor do all plants have flowers or leaves, or even stems and roots. In short there is no simple test as to what is a plant.

Perhaps the plainest statement that can be made is to say that the plant kingdom includes those living things which can manufacture food from the raw materials of earth and air, which seldom have the power to move from place to place, and whose substance is more carbohydrate than protein. But the plant kingdom also includes a host of forms which, like animals, cannot produce their own food by the aid of sunlight and inorganic substance. The best we can say about these is that, apart from their dependent food habits, they are more like plants than they are like animals. Not so satisfactory, of course, but there the matter must rest. Within the plant kingdom are kinds ranging from the single microscopic cell of the bacteria to the giant redwood and the massive, ancient cypress in the vale of Mexico.

One cannot understand either extreme without a knowledge of the other, and particularly not without a knowledge of the mighty pageant of intermediate forms, living and dead. All are a part of the kingdom of plants, and this kingdom cannot be arranged in a convenient row like the successive steps in the solution of some grand mathematical problem. Plants have not moved straight forward from the place and time of beginning. Rather have they developed in many and diverse directions, as the branches from the mighty forking trunk of some ancient tree. Some branches took their way when the tree was young, others when the trunk was tall and old; in that sense are some groups of plants at a higher level than others.

The simple plants of today are not the unchanged precursors of the marvelous orchid and the goodly grape. The former, however, like twigs on the lowest branches of a tree, have remained nearer the ancient level. Life, whether in plants or animals, is not fixed and static. Its expression varies as the generations pass. Often this variation may be induced, or at any rate released, by factors in the surrounding world, as for example, by radium and X-rays. We know, too, that competition is ceaseless, the struggle to survive relentless. And we know as never before the illimitable stretch of geologic time and change through which life has existed. The wonder is, not that our story is not more simple, but that it can be unraveled at all.

We have seen how life, wherever it began, was nurtured in the sea and slowly spread to land, becoming step by step adjusted to conditions there. As this went on, bodily structure and the round of life generally became more and more highly specialized. Functions which within the simplest plant are served by a single cell, were in the newer and more complex types delegated to special cells or groups of cells, like the self-sufficient pioneer homestead dissolving into shops and factories, schools and offices.

As the upshot of such change those two great assemblages, plants and animals, have become so varied and so numerous that the earth is theirs. It is rare to find even a limited spot where some form of life is not equal to the vicissitudes it presents. Salt marshes, hot springs, snow banks and polished tombstones all have their congenial colonists. In accomplishing this magnificent adjustment some 300,000 kinds of plants and more than twice as many species of animals have been developed—not to mention those which have dropped by the way and become extinct. No one will ever score those which have been lost from the ever moving symphony of life. The fossil remains which diligent science has uncovered must be the merest fragment of the missing.

From what has been said we should know that the first task in understanding life is an inventory of its teeming multitudes. Even yet this task is far from complete. The cataloguer of life must name and arrange his material—no easy matter. And by agreement his official seal must rest upon an organism before his scientific colleagues can even discuss it. Within the past two centuries he has very conveniently substituted two-word names, in dead and stable classic language, for the string of descriptive phrases which used to serve as a label for each plant or animal.

Once the organism is pigeonholed it is ready for detailed study. Students of form and structure describe, measure, dissect and compare it by every means that technical practice can devise. Only then does it become possible to investigate its behavior intelligently, applying to the task not only a knowledge of its construction, but lore borrowed from the fields of chemistry and physics. In fact it is the dream of the physiologist to portray what he observes in the cold finality of an equation. With the simpler aspects of behavior he sometimes succeeds; but the living whole remains, in some ways as elusive a mystery as the ultimate particles of matter at the other end of the scale.

If science proceeded always by rigorous logic—which is seldom the case—all of these matters we have been discussing should be disposed of before we could begin to study the part played by a plant or animal in the great web of life. It would be necessary, too, to understand the climate and the soil to interpret its relationship to the physical world. But the battle against the unknown can seldom afford to wait until its road is smooth, and so the student of life and environment—the ecologist—has been obliged to proceed, rough shod for a rough task. Scientists in the fields from which we must borrow so heavily, knowing how far from complete is their own share in the great joint enterprise of interpreting life, are often inclined to be impatient with him. But that will pass.

With these rather general observations about plants and animals at hand, we may confine our attention to the vegetable world. Since this consists, as we have seen, of about one-third of a million species, each with its distinctive character, our task will not be either easy or simple.

One species may agree with another in its requirements for light and water, as do the white pine and hemlock; but as surely will the two differ in other respects. Our problem would be simpler than it is if the differences ended with the species. But species generally consist of varieties, often highly different in their demands and abilities. Golden Bantam sweet corn will be ready for the table in six weeks, while maize from Mexico growing at its side will stretch into the air like Jack's beanstalk and not find time in the three month's growing season of Ohio to put forth tassels. Both are varieties, however, of the same species.

Even within the variety may be found races and pure lines, differing only in minute respects. Yet these differences may be quite significant, affecting for example the number of beans in a pod, or the sugar content of beet roots. And finally comes the individual, always a distinctive entity from other individuals, despite a common inheritance. No two maple trees would fit the same mold, nor even any two leaves from the same tree.

What the individual will be, even if we know his inheritance, is a matter predictable only by the curious, circui-

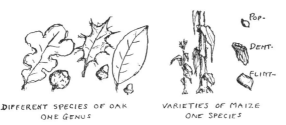

DIFFERENT SPECIES OF OAK
ONE GENUS

VARIETIES OF MAIZE
ONE SPECIES

tous prophecy of statistics. We may be able to predict the limits beyond which he is not likely to vary, and we can even predict his chances of being near the average or either extreme within those limits. But any such predictions must assume that conditions do not change too much.

In fact the effect of changed conditions upon plants adds greatly to the problem of understanding them. Will a certain plant be reasonably plastic under different degrees of illumination, like the red oak? Will it be eliminated by shadow as is the aspen, by strong light like the American beech tree? These are not always easy questions to answer, for perhaps the change from one light intensity to another may conceal a far more significant change, say in moisture supply. It is never safe to consider any one factor without reference to the others. By study of all it is generally found that there is one master factor which may control a given situation. This is often called the limiting factor, for the chain of environment, like that of the proverb, is no stronger than its weakest link.

Some plants are as remarkable for their flexibility as others for the lack of it. The leaf of the tobacco, grown in bright sunlight is harsh and rigid with thick veins—utterly unfit to be used in cigars—hence the square miles of muslin over the tobacco beds in the Connecticut Valley. The cactus, when grown under conditions of heat and drought, will change its whole chemical behavior, storing its carbohydrate food in the form of water-holding gums instead of sugar whose solutions so easily dry out. The wireweed which thrives on beaten paths will quickly die in a gentler situation.

It is possible to divide the plant kingdom roughly into four great groups, each consisting of a number of segments called phyla which are at about the same level of development. Each of these four divisions, while containing very diverse forms, has enough that is distinctive in the behavior of its members to permit us to take a rapid view of the plant kingdom as a whole. The divisions, beginning with the simplest and most ancient, are: (1) the thallus plants; (2) the moss plants; (3) the fern plants; and (4) the seed plants.

The thallus plants have as a rule soft bodies, often minute, generally unspecialized. The moss plants have soft bodies too, but have certain specialized features for living on land, yet with the fern plants require water as an adjunct in their sexual reproduction. The ferns however possess roots, stems, and leaves which the mosses lack. Within these organs are highly specialized structural elements to assist in the conduction and storage of materials, the support of foliage, and the efficient manufacture of food. Because of the presence of such inner conducting and mechanical structure both ferns and seed plants are said to be vascular. But the seed plants have the means, which undoubtedly originated in ferns now extinct, of protecting their juvenile stage within a structure called the seed.

The oldest of these four groups, the *thallus plants,* have by their very simplicity become adjusted to a wide range of conditions. Forms which remain in the ancestral waters of the earth are perhaps the most characteristic. Floating in the plankton of the surface layer or attached to the bottom wherever effective light can penetrate, they are remarkably adjusted to all possible aquatic conditions, supplying forage throughout the sea and lakes. Other types of thallus plant have become adjusted to life upon land, enduring desiccation which would be fatal to specialized plants and making use of scant, occasional moisture with amazing efficiency.

In addition to those thallus plants which utilize sunlight to manufacture food and which are called algae, there are others—the fungi—which do not have this power. Included among the latter are the bacteria. All fungi, with few exceptions, must secure their food from living organisms or from dead organic matter. Decay and disease—both environmental influences of the utmost importance—are the practical consequences of this activity. Certain fungi, living in conjunction with algae, develop a characteristic plant body known as a lichen, found everywhere on tree trunks and rocks. These lichens are important agencies in the formation of soil.

The growth and reproduction of the thallus plants is often sensitively adjusted to conditions of the environment. The group is a perfect museum of the early history of sexuality. Yet their vigorous spread is largely made possible by little reproductive bodies called spores. Some spores swim through the water but many drift about in the air, rising, falling, or moving horizontally as the currents go.

Small wonder then that thallus plants are on hand to grow wherever conditions permit, profoundly affecting the lives of more pretentious organisms in a host of ways. The onset of disease or decay, or the growth of algae in a pool is more often prevented by lack of suitable conditions than by the absence of the appropriate thallus plant. Shutting the lid of a bread box tightly does not bring in the mold; it merely provides moisture whereby the spores already present can develop.

The *moss plants* are low-growing, tufted, and green, requiring water, as we have said, for their sexual reproduction, but producing aërial spores as a consequence of this process. As a rule they are limited to places where their delicate bodies are not likely to dry out. Growing in cushions upon thin pockets of mineral soil if it be moist, they play a humble but necessary part in building soil, and are the precursors of more stately vegetation.

Notable among the mosses are the Sphagnums which grow upon moist, acid soil and upon the acid surface of drying lakes where they often form huge deposits of peat. This is made possible by a feature which mosses share with ferns and seed plants, namely that of more or less continuous growth at the tip. By contrast animals usually grow for a limited time and throughout most of the body.

MOSS, FERN, AND LICHEN

The *fern plants* of today are but a vestige of those which, as we have seen, ruled the earth in the days when the coal measures were laid down but succumbed to the cold and drought of Permian times. Their character can best be understood by noting that the advantages of a highly organized vascular body were in the end outweighed by inadequate protection of two delicate and critical phases: the tender growing leaf tip and the sexual organs. And while the extinct ferns developed the capacity to produce a stem supporting the foliage and connecting it with the root, the stem was seldom able to thicken or branch as it grew older. Well suited to the extensive humid swamps of Carboniferous time, they could not survive more rigorous conditions, particularly in competition with the highly developed seed plants. Today they are largely relegated to life on the humid forest floor where conditions are protected and controlled by the growth of trees, which are seed plants.

As to the *seed plants,* we know that they, too, are of ancient lineage, the older forms being those which, like the pine today, bear their seeds naked upon scales arranged in cones. The flowering plants such as apple and bean, whose seeds are embedded in a fruit, are far younger and more highly specialized. Those which bear cones have generally small, harsh leaves—an advantage under conditions of insufficient moisture and unfriendly habitat. On the other hand the tissues within their bodies are less varied and in consequence less plastic and adaptable in the face of a variety of conditions than are those of the flowering plants. In consequence the latter, when given favorable conditions, are as a rule the more effective.

Moreover, while both kinds of seed plants afford good protection to the critical sexual process and to the infant plant, this adjustment reaches its greater perfection in the group of flowering plants. In both, the delicate male cell is enclosed within a pollen grain and the egg is within the future seed. The meeting of the two is effected in the cone-bearing plants by wind—a biological extravagance which can be appreciated by anyone who has seen the rain of golden dust which covers the water tanks and ditches near the pine forests of New Mexico, for example. On the other hand a large proportion of the flowering plants are able to effect the distribution of their pollen much more efficiently, often having it delivered precisely to the spot through the remarkable behavior pattern of specialized insects. The benefits are not to the plant alone, for the insect usually obtains either nectar or pollen in the process. However it is an interesting fact that some of the more highly developed flowering plants have developed anew the ancestral use of wind pollination. Such are the ragweeds, potent source of hayfever, which will be mentioned again in Chapter XV.

The flowering plants have benefited no less by the enclosure of the seed within a fruit than by protection of the sexual stages. There is an endless variety of fruits, both dry and fleshy, affording the most varied means of transport by wind, water and animals. In this respect, as in structure and organization of the vegetative body, flowering plants truly surpass all other forms of vegetable life. Since their lost beginning in the early Tertiary and their magnificent expansion in the Cretaceous, they have developed in richness and resources until they clothe the earth. Their nearest rivals are the conifers occupying the less genial portions of the earth and the thallus plants within its waters.

Biological success is never without its perils, and with the rise of the flowering plants, enemies and dependents have waxed apace. Notable among these are the fungi and the insects. Yet from both of these groups have come many forms which, as we shall see, form an indispensable part of the picture of life. Without their assistance the present dominating vegetation of the earth could not exist. All are woven into a seamless fabric, each playing its part and becoming adjusted or else passing out of the picture.

No small part of this adjustment is a result of competition among the flowering plants themselves. This competition is not confined to that between different species but exists between members of the same species and even within the individual itself. No one can fail to understand this who has seen the race for light between sister seedlings in a dense stand of aspen or observed the death of the lower branches as the upper intervene between them and the sun. Below ground as well as above, organs of plants compete ceaselessly for the space, the nutrients, and the energy which they require. Earth's vegetation is in a large measure the resultant of this process.

The details exemplifying this brief account of the plant kingdom are beyond the knowledge and comprehension of any single individual. Practical necessity has dictated that certain plants, like the maize, the potato, and the orange tree, should be studied with a thoroughness reminiscent of the zeal with which man has studied his own body. Many of the most interesting and in the end, perhaps the most significant aspects of plant life have been left to the chance curiosity of professors and other amateurs who study plants as a luxurious indulgence with such energy and time as they can spare from the business of earning a living.

Yet slowly and unevenly though it be, our knowledge grows. The Hindu physiologist, using all of the subtle refinements of modern electrical devices, reads into the living plant a sentient character. The realistic American botanist strives to explain the plant as a highly intricate physical and chemical mechanism, expressive of the operation of blind forces. The emergent evolutionist concedes that plant life is a manifestation of physical and chemical laws, but insists that the living individual is a phenomenon representing a new and distinctive level of activity as compared with the inorganic world.

The variety and richness we have attempted to suggest is to the strict mechanist a matter of impersonal, unconscious, mechanical causation. But to the classifier, either of living or extinct plants, the range of form and organization often seems like evidence of an internal principle of perfection, working in the most diverse groups of plants in a fashion which, however imperfect, is yet strangely upward and parallel. With what new concepts or phrases the students of another hundred years will reconcile these divergent facets of truth it is not our privilege to know. It is not even impossible that the human mind may at times weary of the struggle to find unity in the face of diversity and abandon the facts to their own fate, if doing so does not interfere too much with the practical business of living.

However baffled we may be in our search for meaning within the world of plant life, we can be sure of its place in the broader scheme of events. We are quite certain that, in that daily pouring of radiant energy from the sun into this system we call the earth, it is the existence of green plants which arrests for a time the final dissipation of that energy. Moreover we know that the great host of lowly plants which are not green—the fungi—are continually at work, unlocking the chemical treasures of the wasted and the dead, releasing them for new generations to use.

And at the same time that green plants are impounding energy along with materials into food, they are keeping the atmosphere fit for animal life. Much of the waste of life goes back into the air as deadening carbon dioxide. From this plants remove the carbon, releasing the pure and breathable oxygen without which we cannot live. With a neatness truly admirable, this oxygen enables living things to get at the energy in their foodstuffs by breaking them up and carrying their carbon back into the air—carbon dioxide, once more, at the service of green plants. All a bit dizzying, no doubt, like any good merry-go-round; but once aboard, you can count on it to bring you back where you started.

But perhaps enough has been said to show that plants

are much more than familiar, pleasant, useful objects about us. They are indispensable. They are more than a part of our environment, such as it is. They have helped create that environment.

Chapter 10
ANIMALS, LIFE ON THE MOVE

*"I went to the Animal Fair,
The birds and the beasts were there.
The Old Raccoon, by the light of the Moon
Was acombing his auburn hair."*

ALEXANDER the Great, as we all know, took especial pleasure in sending back to old Professor Aristotle all sorts of live animals for his zoo in Athens. And ever since that time generous men, whether millionaires or aldermen, have found great satisfaction in serving as patrons to similar institutions. In their humble zeal to further the cause of science perhaps they forget that a good collection of animals is a swell show, in any man's language. Or it may be they know this.

As for the entire animal kingdom, there we have a wondrous spectacle indeed. Plants are good enough in their way; but, apart from striking and brightly-colored flowers, they represent an acquired taste for the average person. Animals, on the other hand, have an immediate and spontaneous fascination for us, from our cradle up.

Animals, of course, frequently make interesting noises, and generally move about before our eyes. Motion is the key which unlocks the geometry of the world about the growing infant. It creates a sense of space and order in that bewildering mosaic which is his first vision. The magic of motion is with us, waking or sleeping, so long as we are alive. It throbs in our gayest music and creeps into our most somber philosophy. No wonder the moving world of animals bewitches us.

It does not take a great deal of imagination to sense a sort of kinship between ourselves and the higher animals, at least. The bear holds a bottle between his forepaws and pours the contents down his throat like any other toper. The drowsing horse hangs his lower lip half open and looks at the world through partly closed eyes—a perfect type of the amiable, comfortable, old sensualist of our own species. No matter if Cuvier, and later the great Agassiz, looked upon each kind of animal as a distinct and separate production from the workshops of the Almighty;

even children knew better, and the whole body of folklore is there to prove it.

For most of us, this feeling of kinship with the animal world, instead of diminishing, grows with experience. We become aware that animal life is strangely like our own in being based upon a psychic pattern. The grip is clinched when some one animal becomes part of our immediate round of life—the lovable dog, for example, who seems to read our unspoken thoughts and whose behavior is often so much more reasonable, or at any rate, consistent, than our own.

Since, as we have reminded ourselves, both plants and animals live, much that we have said about the one applies to both. Yet the two realms represent two profoundly different trends of expression. The direction of this trend in plants we have considered; to understand its course in animals is our present task. And with all of the good will possible on his part, we cannot expect the scientist to tell us his story unless we share with him some of the tedium of that analysis which is his daily round of life.

Science, unable to define life, tells us that life expresses itself in five great interwoven groups of activity. Four of the five are reasonably familiar to the average man. The fifth, which is called metabolism, is really not so much unknown as we might think, once we get behind its rather oppressive label. The five characteristics of life referred to are *growth, movement, irritability, reproduction,* and *metabolism.*

Both plants and animals may begin life, in fact they

generally do, as a single cell. True, this cell is quite frequently the result of the union of two, but before *growth* can begin, the merger is complete and life continues for the race by starting for the individual. In the simplest forms of organism this new cell is the individual and its round of life is completed by dividing to form two new individuals quite like itself. This process continues indefinitely, now and then interspersed with the sexual fusion of two cells. In some simple forms, however, there is no trace of sexuality.

In more complex organisms when the initial cell divides, its daughter cells cling together according to an inherent pattern, the mass ultimately assuming the form of the mature individual, and a starfish, a snake, or a man emerges. In the higher plants, as we have seen, this process of growth continues throughout the term of life, in the buds, the root tips, and beneath the bark, by the division and subsequent transformation of cells. But in the higher animals the mass of dividing cells rapidly establishes the form of a juvenile individual, then becomes adult more by enlargement than by change of shape. In other words the chick and pup look more like the hen and dog than the newly sprouted acorn resembles the old oak. The embryology of the plant endures throughout life; that of the animal is early completed. Thus it is, since growing things are in themselves delicate, that the plant kingdom abounds in protective structures such as bud scales and bark, while the life history of animals exhibits many means of protection of the tender young, both before and after birth. And because of the ineffectiveness of such protection the once great groups of ferns and reptiles are now mere biological curiosities, no longer powerful themes in the symphony of life.

In *movement,* no less than in growth, do animals show their character. Here, too, the lower plants and animals differ less than the higher, for many simple plants have the power of locomotion, a power which may be retained by certain cells even in the highest plants. But in the latter the necessity for permanent anchorage in the soil—source of minerals and water—makes for a rigid framework. And the material is available in the form of surplus carbohydrate, chief product of the green plant. Each living cell is boxed within a wall of cellulose. The power of locomotion is lost; movement is by means of growth alone.

Animals, on the other hand, have retained and extended the power of locomotion through the early development of a muscular system, largely protein in character. The necessity is obvious since animals are dependent upon the anchored plants for food and must go where this food is. Without great powers of locomotion the animal kingdom could not long survive.

Since this power involves the need for accurate coördination, it would be impossible without the development of a nervous system which not only transmits impulses but adjusts them to experience. This is not to say that plants are utterly lacking in coördination or in the transmission of impulses; but these activities reach no such perfection in them as in animals.

The capacity to transmit and to record experience throughout the organism is called *irritability,* a word of broader meaning in science than in common speech. It is the third of the five great phases of activity which we are discussing. Irritability is an essential means of guiding the movements of animals through the perilous maze which separates benefit from harm. Moreover in animals, as in plants, it is the means by which internal processes are maintained in as much harmony as possible with the external world. The trout heads upstream because he is more comfortable so. The timorous husband allays a slight sensation of intestinal unrest by recalling the precise details of his wife's parting injunction.

Irritability in animals culminates in the development of highly perfected sense organs which receive impressions from the environment, and in a central nervous system whose most complex expression is the human brain. By these agencies experience of the outer world is received, recorded, compared, and used as a basis for future patterns of conduct, no less than of immediate response. Thus behavior is conditioned, not only by the experience of the present, but by the accumulated experience of the past. This accumulation reaches back through the life of the individual and, in ways not understood, back through the life of the group to which he belongs.

Nowhere is this more amazingly manifest than in that next great phase of behavior on our list, namely, *reproduction.* By this means the stream of life continues, finding its expression in new individuals. In the simpler forms of life this is accomplished by division or separation of the old individual into parts. Such a method is prodigiously efficient, as the rapid growth of yeast and certain disease bacteria proves. Even in some more highly organized species this process is successful. A few worms, a pair of scis-

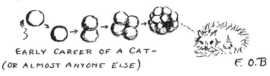

EARLY CAREER OF A CAT—
(OR ALMOST ANYONE ELSE)

F. O. B

sors, and a box of good earth will set one up in the fishworm business; and all of the navel orange trees in California are parts of an old tree which once grew at Bahia in Brazil.

But for the most part reproduction involves the phenomenon of sex with an intricate complex of attendant behavior. Two cells, each produced in the bodies of separate individuals, must fuse to establish the new individuality. This becomes possible through precise adjustments of adult form and action of the parent individuals, not only toward each other but toward the rest of the environment. There is scarcely a phase of animal experience that might not be permeated by sex; whether it is in fact so influenced is a matter for scientific investigation. In the meantime it is well not to underestimate the force of the groundswell of sex, or to think that it can be excised from the rest of living.

In many insects, for example, the highly specialized adult winged stage finds no other expression—not even the search for food—than that of reproduction. The May fly has no functional mouthparts. It emerges from the water, mates, drops its eggs, and dies. Even the butterfly is the dilettante she appears to be in her casual sipping of nectar; but she has a grim persistence when it comes to laying her eggs upon a proper source of food for her young, and her mate an equal persistence in finding her before that operation takes place.

In the sexually adult vertebrate animal existence is an end in its own right, usually prolonged, absorbing, and effective, yet permeated from beginning to end by the deep-seated cycle of reproductive behavior. The inception of a new human being may not literally be a gleam in its father's eye, but that is close enough to the truth for practical purposes. And infancy terminates, by civilized law, at somewhere near twenty years. Human parentage spanning this long gap, meanwhile attempting to live life for its own sake is at a far pole from the dancing, dying May fly with a few hours' task neatly cut out. And if civilization survives material disaster caused by exploitation of the earth, she is still not safe from the growing duality of spirit, the undisciplined confusion which assails her individuals during their years of reproductive maturity. No animal can survive the destruction of its nesting habits, not even man.

Enough has been said to show that the work of reproduction in animals goes far beyond the development of structural and mechanical devices such as are present in plants. In both realms, reproduction gives rise to problems caused by population growth with consequent pressure upon the means of survival. This appropriately introduces the last of our five phases of behavior, that of metabolism.

The word *metabolism* is used to include all of the building up and tearing down of chemical compounds within the living body, along with the accompanying storage and release of energy. It is less familiar to the average person than the words which describe the other phases of behavior, because there was no means of understanding metabolism until modern chemistry and physics had shown the way. Anyone can observe the processes of growth, movement, irritability, and reproduction; they are the common and accepted tests of life. But to observe chemical change and its significance is a technical problem. Yet, curiously enough, now that we have the means to understand it, we find that it is in many ways a simpler phase of activity than the others. For them it provides the structural material and the energy, just as technology provides these necessities for modern civilization. And in the same way, it is easier to comprehend the work of the engineer than the consequences of his work in terms of economics, politics, and morals—in other words the effect of his work upon human living.

As we have said, metabolism includes both constructive and destructive change. Until the adult condition is reached, there must be a net gain in the material and energy content of the organism. Senescence is marked by a net decline, and death by an end of orderly metabolism. Apart from the fact that plants can manufacture food within their bodies while animals cannot, the metabolism of the two realms of life has more in common than is generally realized. Both plants and animals use food, largely the same food, as a source of material and energy. In both, food can enter the living cells only in solution, save for some very simple, microscopic animals whose cells surround the food they use.

The typical animal body, unlike that of plants, is so constructed that solid as well as liquid food can be ingested and prepared for use. After ingestion, such food is worked upon by digestive ferments and the soluble products are absorbed, the undigested remnant evacuated. The soluble nutrients thus obtained are carried throughout the body and so made available to the living cells. In the work of the living cells, byproducts are produced, sometimes poisonous in character. These are carried to the lungs, kidneys, and skin, there to be excreted or thrown off. In this process the lungs excrete the carbon dioxide and water from respiration, the kidneys attend to nitrogenous waste from the breakdown of protein materials, and the skin gives off, among other things, water and salts.

The material interchange between animal and outer world is a much more involved process than the simple absorption of dissolved minerals and the exchange of gases by the green plants. Little wonder, then, that the waste products of the animal kingdom are such an important source of concentrated organic material for the soil. And quite as little wonder, considering the incessant internal circulation needed to maintain animal existence, and the locomotion necessary to obtain food and to reproduce, that the world of animal life is an extravagant user of the energy which plants accumulate.

So far as there is a key to the intricate problems of animals and environment, it lies in food. Animals must eat. Ingestion is the first, most essential step. Unquestionably animal behavior can be affected greatly by other factors than food, for example, by temperature, acidity, light, moisture, gravity, and pressure. Moths which fly at twilight (crepuscular is the musical adjective which describes them) will invariably cluster in the middle of a long glass tube grading from opaque at one end to transparent at the other. The gopher seeks his burrow and becomes inactive as the thermometer drops below about 38° Fahrenheit. Countless other examples of the control exercised over animals by purely physical circumstances can be cited. And yet the major logic of animal behavior lies in the distribution of the food supply.

Food is sometimes defined as any material taken into the body to sustain life—thus including water, oxygen, and salts. But these alone can never sustain animal life. For that purpose organic material must be at hand, and as we have reiterated, its ultimate source is plant life. Plant material may be, and often is, reworked by other organisms before it is used by a given animal. This sequence is known as the food chain, inspiration of the rhyme on an early page of the preceding chapter.

It is not enough for an animal to be present where his food is, although that is the first step in securing it. He must possess features of physical structure so that he may utilize the food. If it does not present itself before him, as food does to the coral and to the indolent bottom feeders of the ocean, he must be ready to seek and pursue it.

If food is not in the form of fine fragments such as the tiny organisms which the whale, leviathan that he is, filters from sea water, the animal must have a mouth equipped to dissect or even to grind it. Thus do the caterpillars which defoliate our trees, or the deer which browse upon the foliage of the forest. Insects, for example, which do most of their feeding in the larval stage, are divided into those which possess biting mouthparts and those which are able to pierce and suck the juices from their sources of food. The grasshopper chews while the plant louse and leaf hopper, like the mosquito, penetrate. Horses, antelopes, and cattle have mouths marvelously adjusted to the use of grass and other herbs, and digestive systems which are no less so. The snake, thanks to a double hinge on his jaw, can accomplish prodigious feats of bolting.

Structural adjustments are by no means limited to those suitable for seizing and ingesting food. Pelican and squirrel both have pouchlike structures which enable them to gather more than is immediately needed and convey it away, the one in his throat, the other in his flexible, furry little cheeks. Bees and their flower-visiting kin not only ingest and transform the watery nectar into honey for future use, but pack huge loads of nutritious pollen—rich in protein and carbohydrate—into the basketlike bristles upon their legs.

Dentition—the pattern and construction of the teeth of vertebrate animals—is one of the surest guides to identity because tooth character is so intimately related to food habit. The chisel-shaped biting front teeth of the grass eaters are in striking contrast to the sharp, tearing teeth of the animals whose victims they so often are.

Frequently indeed the entire body, or at any rate organs remote from the digestive system have an important structural relation to food habit. Woodpecker and kingfisher are not dissimilar birds in general appearance. Yet the former, with his powerful cutting beak, and his strong, stubby tail which serves as a brace, is more steeple jack than flier. His mortality since the coming of the auto has been terrific, thanks to his slow and clumsy flight. The kingfisher, on the other hand, is a powerful, graceful flier and diver; even his crest feathers seem to be streamlined. And in place of the insects which the woodpecker drills out from trees, the kingfisher lives on fish by outdoing them in their own element. Likewise the respective food habits of giraffe, horse, and wolf are clearly reflected in their skeleton and musculature. The paleontologist seldom hesitates, in the presence of some ancient mammalian or reptilian skeleton which can only be clothed in flesh by the imagination, to tell us with assurance how the animal made its living.

The adjustment between food and feeder—often so perfect that it seems prearranged—goes beyond the matter of structure. Naturally it permeates behavior, from matters of digestive chemistry of which the animal is quite unaware through to the most complex features of his psychic pattern. Men who have successfully tamed coyotes and developed considerable attachment between themselves and these wildlings all sadly agree that their pets cannot be trusted as guardians of the poultry yard. Bird migration, astounding circumstance in so many ways, is sensitively and usually with effectiveness related to food supply.

The story of feeding is inseparable from that of reproduction. Nowhere is this more strikingly shown than in the insects—specialized as they are for prolific increase. Frequently the food habits, if any, of the adult which breeds and deposits eggs are entirely different from those of its offspring, the juvenile larva. The ephemeral May fly already mentioned shows this perfectly, and of course so do moths and butterflies, when compared with their caterpillar stage. Clearly there are few more interesting or difficult problems in biology than the explanation of such elaborate adjustments.

Behavior in animals is not, of course, limited to the individual or the pair which mate. It extends to the community, which in its turn is no less influenced by the problem of nutrition than is the individual. In the remarkable colonies of bees and ants there is the highest degree of specialization. Reproduction is the task of a minority of the individuals, the rest functioning almost mechanically for the protection, care, and feeding of the immature. The breeding, feeding, and migration habits of social birds, notably waterfowl, again illustrate coördination of individual and group. To imitate, in a squadron of airplanes, the perfectly timed maneuvers of a flock of canvasback or mallards on the wing requires the most exacting discipline and perfection of mechanical resources. Which is to suggest that group adjustments in living animals are not a simple matter.

In stressing the importance, even the predominance, of the food problem in relation to animal adjustments we cannot, of course, think of it apart from the rest of the environment. It simply is, more often than not, the limiting or master factor. Yet of almost equal importance is the problem of physical safety which looms so large in a world of enemies and competitors. In the world of nature there is no such thing as complete security. The problem becomes one of living with enemies and rivals rather than one of securing freedom from them. The remarkable military formations of a grazing bison herd in the face of wolves or other danger represents a measure, not of absolute, but of effective security. No living wall of cows and heifers, flanked by a cordon of powerful bulls, will keep the predator from snatching the weak, the newborn, or the stragglers as these are left behind by the moving herd. Yet by and large, the living wall worked well enough to enable the bison to survive until he met the mechanized and ruthless attack of the white man.

Abolitionists and violent reformers would do well to ponder the workings of nature, particularly the principle just indicated: that security and protection are effective, not absolute matters in life. The essential problem of human slavery was not seriously modified by the military conquest of a group of plantation owners. A city in which vice is segregated is likely to be, on the whole, a safer place to bring up boys and girls than one of the same size in which vice has been "abolished." And the rampant evil of alcohol is more likely to submit to the powerful harness of social convention than to the whip of suppression.

The seriousness of a biological enemy is not always in proportion to its size. The microörganisms which produce disease are of frightful aggregate power. Yet the fire of experience often produces a measure of immunity, or at least tolerance toward them. Or the menace may be met less directly: the victim may move away from or avoid those conditions most favorable to his microscopic enemies. The white man, until he learned the proper sanitary régime, had thus to avoid the malarial and fever-ridden tropics.

Nor can we neglect, by any means, the constant problem of protection against, not the living, but the inanimate forces of nature. It so happens that by virtue of suitable skin-covering and a remarkably sensitive heat balance, many animals are able to endure a wide range of climatic factors. Even so there are rather definite limits to which each animal species is best adjusted. Man and his dog have done somewhat better in the matter of such limits than most of their fellow creatures, but that has been largely by grace of cultural achievements. Their success merely emphasizes the general truth.

Here, as in the adjustment to food, the time-pattern of animal behavior is extremely important. Periods of inactivity, whether in egg, embryo, or dormant adult, are generally timed to afford, not perfect, but effective adjustment to unfavorable conditions. The blossoming willows of early spring, with their rich freight of pollen, do not call out the bees directly. But the same vernal sunshine which arouses the one releases the other from the bondage of the cold and permits him to fly forth.

The lengthening days and early warmth which stimulate the northward flight of the robin from the Carolinas to Ohio sometimes lure him to death by starvation or unseasonable frost. But more often they do not. At any rate there are robins in plenty left. The cold-blooded snake, slowly stiffening in the cooling air of autumn, finds pockets of warmer air remaining in the recesses and caverns of the rock, and seeks them. As these in turn cool he enters upon his winter torpor, with the advantage, however, of the greatest protection possible against storms and exposure.

The fat of polar animals seems to have a lower melting point than that of their tropical kin. It may be thus a better source of thermal energy at the usual temperatures these animals encounter, and it certainly is a more plastic body substance because of this property. Moreover, this fat has been obtained from the green sea plants which flourish during the brief but intense arctic summer and is rich in the sun-made vitamins so essential to normal nutrition. It serves therefore as the reservoir which keeps in health the animals dependent upon it during the long arctic night.

One cannot hope in such a brief discussion to do more than draw the broadest principles and to select a few striking examples from the hundreds of thousands of varied species of the animal kingdom. The separate groups themselves are known in detail only to specialists, and frequently these men are so concerned with the study of structure and classification that they can pay only the most cursory attention to problems of environment. As a practical working expedient, the student of animals and environment analyzes the population of an area and determines its general relationships, but refers the identity of his material to various specialists. So great is the number of animals, and so flexible their adjustments to all sorts of conditions, that it is scarcely possible to stamp any particular group in terms of its ecology.

Conveniently the animal kingdom may be divided into the invertebrates, more ancient, and the vertebrates, more recent, but by no means a youthful group. Both flourished first in the sea, and there the older group has been compelled to share its dominion with those remarkably successful vertebrates, the fishes. Yet the marine invertebrates, one-celled protozoa, many-celled holothurians, worms, molluscs, and crustaceans are a mighty host. Eating marine plants and each other, they are eaten by fish. Nevertheless not all invertebrates are victims of the fish, many having become highly successful as parasites and other dependent forms of life.

Moreover the invertebrates have spread to land with no small success—success enough indeed to justify the melancholy witticism that "the last thing on earth will be a living insect on a dead weed." Here on terra firma a modest proportion of them have survived, as have the crayfish, certain worms, and fresh-water sponges, by their conservatism. Hazarding little in the way of biological innovation, they remain adjusted to the same general diet and round of life as their ancient forebears.

But for the most part, the dazzling success of the land invertebrates has come through their remarkable evolutionary flexibility. They have changed in structure and habit as the higher plants and animals evolved, at every step tapping the new and rich resources afforded by their more pretentious victims. So obvious has this multiplication of dependence become that it took no trained biologist to write Swift's lines of fleas upon fleas upon fleas. More than a score of insects cause galls upon the hackberry, perhaps as many upon the rose, and many more upon the common oaks. The veterinarian must reckon, besides bacterial diseases and nutritional disorders, with bots, screw worms, ticks, lice, fleas, tapeworms and a host of other troublesome small animals which afflict domestic livestock. In this facile, if servile, adjustment to the increase of higher land life, the invertebrates resemble the fungi, prime sources of disease in plants.

Yet it would be a distortion of that superb symphony which is all life to imagine that the persistent success of the lowly invertebrates has been wholly at the expense of the newer and more specialized forms of life. We have already noted the importance of insects in the pollination and dissemination of the higher plants. Many processes which are beneficial, even essential, to the higher organisms can be attributed to the invertebrates and to the fungi. This is especially true with respect to the material cycle of the soil. We have seen how essential are worms, insects, and other small animals to the breaking up of organic waste.

Even those forms which are sources of disease and injury are not completely discordant elements. Observation shows that in a state approaching a balance, such as we find in nature undisturbed, the injurious forms are only moderately active and destructive. Borers which infest living trees, grasshoppers and rodents such as rabbits and gophers, apparently become much more numerous under conditions induced by man than they otherwise might be. Man likes to kill weasels; weasels are very fond of young gopher meat. In virgin conditions, the victims of parasites and predators are very often organisms growing in unsuitable locations or deficient for some reason in normal vitality. Moreover, the means of subsistence upon earth are limited, and the speedy price of universal survival would be universal death.

Finally, the vertebrates after emerging from the sea as amphibians and slowly advancing through multifarious reptilian forms of which the birds are a direct offshoot, have culminated in the higher mammals. Marked by a superior nervous system and increased efficiency in protection of the unborn young, the land mammals have specialized in keeping with the rapidly enriching land flora. As among the great reptiles which preceded them, the earlier forms to develop fed upon plants. Subsequently came the carnivores which fed upon these vegetarians.

Excepting for the continent of Australia, where no large native mammals of modern type existed, the grazing, browsing, and carnivorous placental mammals were in possession of the earth until displaced or subjugated by

CAMP FOLLOWERS
DRAWN—LIKE ALL THE PICTURES IN THIS BOOK—
WITHOUT MODELS AND WITHOUT REMORSE!

man. Largely from the vertebrates about him, man has selected his domestic animals. Thus incorporated into his culture pattern, the problem of sustaining them under the protection which he supplies is not the least difficult in the creation of a new equilibrium of nature.

And in addition to the animals which he willingly sponsors, man's communities are everywhere thronged with camp followers from among the animal kingdom. Among the invertebrates the housefly, many body parasites, and the cockroach are familiar examples. Among the vertebrates the English sparrow, the barn swallow, the brown rat, and the mouse are no less well known. Occasionally welcome, often indifferent, and still more frequently destructive and predacious, these uninvited guests serve to complicate the problem which man has called about his head by proclaiming himself lord of creation.

Chapter 11

THE LIVING ENVIRONMENT

"STONES ain't people," said the homesick little city girl, sent by kindly friends to the mountains for the good it would do her. Thus neatly, if unconsciously, she divided her environment quite as the scientist does, into the physical and the living. And of the two, by far the more significant to her was the portion which included her own fellows. This is true of most of us—often to the extent that we quite forget the importance of our physical surroundings.

For any plant or animal, high or low, the presence of its own kind is a most immediate and effective influence. But the living environment also includes large numbers of other organisms with which the individual living being must come to terms if it is to survive. And the range of relationships thus established is bewildering.

No animal or plant lives unto itself. We may bring it into the laboratory to study it and for the time being ignore this great principle of nature, as a matter of convenience. But in the end we are obliged to take the knowledge so gained back into the field where our specimen lives and use it to explain the very complex processes we see going on there. The ultimate and visible expression of these relationships in the living world, as we shall see, is the formation of communities. But to understand communities we must first attend to the mechanisms which operate between and among living things.

Among the lower forms of life, a single individual can often procreate its own kind. A single yeast plant, placed upon a disc of sweetened jelly, divides rapidly to form new individuals, as we may see with the microscope. Yet these individuals cling together in a colony whose fortunes illustrate very precisely the sentence with which the preceding paragraph begins. Each successive division of cells doubles the population, but we find in practice that this process soon comes to an end.

One not acquainted with the business of handling small organisms can appreciate the situation if he were to imagine the population of his own city doubling overnight in number of adults. Stores, transportation, shelter, and utilities—even perhaps the water and sewage supplies—would be taxed beyond their limit. Some adjustment might be made for a time, particularly in the midst of an efficient and wealthy state, with ample reserves. But suppose the same thing were to happen the second night, and once more the third. Inevitably the smoothest organization would break down. Hungry people would be at each other's throats, battling for the sheerest necessities. By migration, starvation, or bloodshed the surplus would be cut down. One can even imagine the hardiest standing about with clubs ready to strike down each newcomer as he appears. Somehow, if any are to survive, a balance must be established.

Returning to our yeast colony which may actually double the number of individuals thrice in an hour, there must be ample food in suitable form to take care of this increase. Growth produces not merely physical crowding and pressure upon the food supply, but an accumulation of waste products, quite generally poisonous to the growing organisms. These poisons are most concentrated near the center of the growing mass, which is also most remote from the surrounding food supply. The process of reproduction is sensitively adjusted, not only to the amount of available food, but to its quality. Inevitably, even though growth continues vigorously at the edges, it must slow down in the central, oldest portion of the colony. And in the end it must slow down, even cease, throughout the entire group.

To prolong the process indefinitely the food must be continuously renewed, foreign organisms kept out, and the surplus population got out of the way. In the laboratory this is done by frequently transferring a tiny fragment of the colony to fresh, clean, nutrient material. By such means cultures of cells can be kept going and even made to develop into living plants. But nature is not so considerate and the population which grows by what it feeds upon must eventually, like the human wayfarer, reckon with the host.

Thus inseparably, even in the simplest forms of life, is the fate of the individual bound up with the presence and activity of its fellows. And though every member of the

COLONY OF YEAST   CLUMP OF IRIS

colony may begin existence, as it does, with the same inherited tendencies, yet the position it has within the colony will affect its chance to be nourished, to avoid toxic wastes, to grow to maturity, and to reproduce. It cannot escape the effects brought about by the living environment of its own kind.

Reproduction is seldom such a simple matter for organisms which stand higher in the scale. Yet many of them, particularly in the plant kingdom, do have nonsexual methods of increase—such as the bulbs of the multiplier onion, or the underground stems of the iris. And anyone who has grown irises for a few years knows that a clump of them will behave very like the colony of yeast we have described. The older central portion will become crowded and dwarfed while the margin remains vigorous; so that the careful gardener is forced from time to time to end this fraternal warfare by breaking up the clumps and spacing them in the ground for a fresh start, discarding the surplus as he does so.

Nonsexual reproduction by spores—cells set apart and often produced in great numbers—also occurs in the higher plants, it is true. But it has become virtually a part of the sexual rhythm. The plants so produced are not like the parent; instead they are microcosms concealed within the cones or flowers, nourished there and functioning only to produce the sexual cells. For practical purposes, the higher plants, like the higher animals, are essentially sexual creatures.

To sexual individuals, the presence of their own kind of fellow beings is a matter of the most profound importance. Without a source for both kinds of sex cells, the means of bringing them together, and of nourishing the young until they can fend for themselves, the species must die. There are a few animals like the earthworm, which produce both the male sperm and the female egg within one individual. But fertilization is accomplished by the mating of two animals, one fertilizing the eggs of the other.

And while, in plants, various divisions of the sexual function are found, by far the greatest number of flowering plants combine the two sexes in one individual. The typical flower is "perfect," having both male and female structures. In such cases self-fertilization may occur, being the rule in wheat, pea, and violet; but more often it does not. A common reason it does not is the phenomenon of self-sterility—incompatibility of sex cells from the same parent. But in addition many plants have elaborate mechanical features which minimize selfing and promote crosses with other individuals of the same species. The two sets of sex cells may ripen at different times, or be shielded from each other. Thus there is a constant admixture of new hereditary material, making possible new combinations and consequent variety.

In the very act of inception of the new generation there is involved a sensitive adjustment between two living organisms. Among the higher animals, as we have already had occasion to note, this adjustment is carried to a much greater degree than among plants. Structural, sensory, nutritional, and complex psychic features are involved. It is a matter of common knowledge that the reproductive urge in animals may equal or even exceed that of self-preservation. This is notably but not exclusively true in the female. The male spider, for instance, marches to certain death at the jaws of his voracious mate as recklessly as the mother of a litter of terriers will hurl herself against overwhelming odds in defense of her brood.

Sexual reproduction, while its effect may be more deliberate, in the end results in population pressure and competition, quite like that we observed in the yeast colony. In one respect, however, it is different. The individuals produced by it have at least slight hereditary differences. And these differences, in the struggle for existence, may be quite as important as the order of birth and position in the community. Swiftness, good vision, vigor, height, tol-

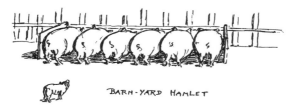

BARN-YARD HAMLET

erance of crowding or of peculiar chemical conditions may vary widely in the same set of offspring, and may mean the difference between life and death.

Competition within the species is always severe, because the various individuals all require the same things. Its practical importance is very great. The grower of fruits, cereals, and livestock regulates it with care in order to secure the best returns. The forester must understand it; an area of lodgepole pine thirty years old may contain as many as sixty individuals in one hundred square feet, while the same plot may show only five or ten surviving a century later. The quality of the timber may be greatly improved by artificial thinning. And the fact of human competition weaves itself deeply into any realistic social philosophy.

Competition involving different species is keenest, of course, between those whose requirements are most alike, but extends to all which share some common need. This may be the result of a similar root habit in plants, which is why the sunflower is a costly weed in the cornfield. Or competition may result from a common weakness for freshly dug earthworms, such as is shared by robin and sparrow. But back of this, the different species differ in other respects, sometimes throwing the advantage one way, sometimes the other. The robin, for example, will dig his own worms; the sparrow much prefers to steal them from the industrious robin.

In the competition between species, man has wielded a powerful influence, more often effective than intelligent. When the bison and longhorn were competing for the western range, the issue was settled by the Sharps rifle. And weeds, which the farmer accuses of driving out his crops, are most frequently the product of his own mismanagement, the breaking of cover on more ground than can be properly handled.

Other outsiders than man may influence the result of competition. There is, in fact, a great array of possible kinds and degrees of relationships among living things. Relationships may be accessory, incidental, dependent, mutual, beneficial, injurious, or lethal—to select but a few words from the list which describes them. Here, as everywhere in the realm of biology, it is the genius of facts to intergrade and run into complexity. With more than a million known species of plants and animals, it could not be otherwise.

The deep-seated, mutual interdependence of the plant and animal kingdoms has already been called the basic principle in ecology. But in addition the records show many instances of special interdependence, some in the same kingdom, others not. One of the most significant and interesting of these exists between the legumes (beans, peas, clovers, etc.) and certain bacteria which grow upon their roots. To manufacture the proteins with which their seeds are so richly stored the legumes must have an ample supply of nitrogen compounds. The bacteria, to grow, must have sugar in abundance, and they get it from the roots of their host, whose leaves make a surplus. In turn the bacteria utilize the energy so obtained to fix nitrogen from the soil atmosphere into compounds which the roots can absorb and pass upward, to be made into protein. Of course the legumes, given an ample supply of nitrogen compounds in the soil, can thrive without the bacteria; and the latter can be grown in test tubes, with artfully contrived nutrient solutions. But under the ordinary conditions of nature each thrives best in combination with the other.

An equally striking case is furnished by the termites, so-called "white ants" which are really not ants at all, but members of a much more primitive group of insects. These terrors of the home owner in warm regions will burrow into wood wherever it touches the ground, or even make clay tunnels up to it if necessary. They cannot live in the light. But the important thing for us to note is that they live on wood, digesting and assimilating it. Yet, curiously enough, it has been found that when termite eggs are hatched in a sterile container and the young fed on sterilized wood, they starve to death. For the normal termite contains microörganisms in his digestive tract which enter his body early in life and which are essential to the digestion of wood. Unless he becomes infected with them, his chances of surviving on his usual diet of sawdust are nil. Which is a reminder that we ourselves are walking botanical gardens, and that it makes considerable difference in our health what kind of a flora we carry about in our digestive tracts.

Often it is difficult to decide correctly about the degrees of benefit in a given relationship. On the whole the adjustment between flowering plants and the insects which pollinate them is reasonably equitable. The use of insects instead of wind is so great an economy of pollen that the commission collected by the useful little animal in the form of nectar or pollen itself is insignificant.

On the other hand we have the lichens whose predominantly gray-green, but often brilliant colors clothe rocks and tree trunks nearly everywhere. These are dual organisms—green algae caught in a weft of fungus threads—and the whole shaped to a very specific pattern. The algae make the food, no question about that. Reproduction occurs by the fungus infecting and entrapping free algae, or by bits of the combined mass blowing about and lodging. Probably the combination grows in many places where the alga could not possibly live without the fungus threads which hold it enslaved. With our own prejudices we call the arrangement an unequal one, but there are arguments both ways. The fungus is a parasite upon the alga, yet both are members of a balanced, effective, and highly interesting system. Perhaps this will console us for the vastly different degrees of ability we possess, and the varied rôles we fill in society.

The degree of dependence of one organism upon another varies so much that we are warranted in thinking the more extreme types of parasitism to have developed by the gradual loss of independence. Families of vines, which depend upon other plants for mechanical support, are likely to have close relatives which have gone a step farther and sent roots into the living tissues of their host. Among the wasps, which as a group are savage predators, we find species which are content to share the food provided by more energetic insects. We call these commensials—uninvited, but apparently unresented poor relations. And closely akin are not a few downright parasites, notably forms which deposit their eggs within the bodies of insect larvae, nicely balancing the time of hatching with the maximum food content of their overstuffed victims.

To many of these parasitic wasps we have, as human beings, good reason to be grateful. We may scorn their degenerate mode of existence, but we do not scorn to send our most erudite men of science to the ends of the earth in search of them when we need their services as allies in our warfare against injurious insects.

The fungi, second largest of the groups of plants, are with few exceptions dependent forms, obtaining their food from either dead or living organic stuff. But it is notable that they carry clear marks of relationship to the algae which are green and able to nourish themselves. Perhaps most interesting and suggestive, in this connection, are those forms which, like the Euglena, can function as green plants in the sunlight, or carry on in darkness if suitable organic food is present. In the latter case they are essentially fungous in character.

No less remarkable than its origin by "degeneration" from independent organisms, is the precise, often elaborate, adjustment of the parasite to its host's peculiarities. No sycophant, intuitive to read the eyebrows of a tyrant, or the corners of his mouth, is more delicately geared to his means of livelihood. And the reason seems fairly plain —the stakes are life or death. It is even fatal for a parasite to become too efficient and thorough in its work of destruction. The bracket fungus which cannot infect a living tree except through a wound or fire scar bids fair to last longer in North America than the chestnut bark blight or white pine blister rust which can enter and kill any host upon which their spores may light. Yet the chestnut disease, which has virtually wiped out the native American chestnut on our eastern seaboard, is a fixture in Asia, where the native chestnuts have considerable resistance to it.

The qualifications of a parasite must necessarily be complex in some respects, even though greatly simplified in others. In the first place its life history must be timed to that of its host. It gains access when its host is most defenseless and grows when its host most abounds in nutriment. It must have structural peculiarities nicely adapted to its victim. The germ tube of the rust spore which enters the pores of the wheat leaf, the hairlike ovipositor of the ichneumon fly which can penetrate an inch of oak wood and lay its eggs in a borer larva, not to mention the pestiferous drill of the mosquito, are all examples of this fitness.

Moreover, such structures are useless without an appropriate type of behavior to accompany them. The germ tube grows toward and into the pore, not away from it. The ichneumon fly can locate the hidden larva with utmost precision—a task that would tax a good microphone in the hands of an expert physicist. And the mosquito needs no light to locate his unhappy victim, for he is guided unerringly by the odor of human perspiration—a fact which greatly assists him when that victim begins to fret and fume at his unwelcome attentions.

Essential, too, is the chemical relation between parasite and host. The former must be able to digest his purloined food, of course. This means an appropriate battery of digestive juices, or enzymes. The termite, as we have seen, cannot produce these himself, but must have them furnished by the microörganisms which infest his digestive system. The fungus which enters a plant and lives upon its carbohydrates, fats, or proteins, must be able to dissolve the cell wall which contains them, in addition to digesting the food itself. The insects which produce galls, or swellings, whereby their larvae are at once housed and fed with plant tissue, go even farther. For the chemical emanations of the larva stimulate the growing plant tissue, changing the whole course of its growth and producing galls of very definite form, composition, and texture. Thus the insect, by chemical means, assumes control of the processes of the plant, to its own advantage.

In addition to the more purely nutritive features of chemistry, there is always the deeper matter of compatibility. So far as we know this is a matter which depends largely upon the proteins of host and parasite. The proteins themselves are complex and highly variable—so much so that chemists calculate that each kind of plant and animal could have a chemically distinct protein, and probably does. Victims of asthma, hay fever, and food rash need not be reminded of the often serious effects of foreign protein on themselves, for such proteins from pollen, molds, feathers, food and other sources are the cause of their particular troubles.

The parasite, practically speaking, is barred from any effective attack upon a host whose proteins are to him incompatible. Or if the host is so sensitive to his presence that any cell invaded dies, the parasite may become sealed as in a tomb of dead host tissue and be unable to grow or thrive. Naturally in most cases of parasitism there is at least a degree of mutual tolerance. In a subsequent chapter we shall consider disease, often the result of poisoning due to the presence of the parasite, once he becomes well established.

We have sketched some of the more striking effects of one organism upon another. But to get the real import of these so-called biotic influences we must view what, for lack of a better term, we may call the balance of nature. Even where man has upset the delicate balance which exists in his absence, we see the pattern of life manifest as an equilibrium of sorts, developed among living things working under given conditions of soil and climate. The frame of reference, then, is afforded by the physical factors of the environment. Within that frame the struggle and adjustments of living things create a pattern.

To produce a forest, for example, there must be sufficient moisture in soil and air. The temperature and growing season must be suited to the requirements of trees. The seeds of trees must have means of access and growth, both of which may depend upon other organisms, perhaps rodents and fungi. But once the forest has developed, it dominates the life of the area. It shades the ground, so that only shade tolerant species can form the undergrowth. It lessens evaporation, so that these tolerant forms are supplied with humid air. If deciduous, it permits light enough for herbs to grow and flower in the early spring, before its own leaves appear. The accumulation of its leaves and twigs furnishes an abundance of humus for saprophytes and subterranean insects, worms, and other invertebrates. Its nuts and fruits furnish food for characteristic animals.

Birds nest in its branches. Browsing animals seek its shade and protection, meanwhile feeding upon its lower foliage and twigs. As its animal population multiplies, carnivores enter, both for food and concealment. Even in maturity and death, the trunks afford food and protection to fungi and insects. The natural history of a fallen tree is a graphic panorama of successive communities of small organisms. And finally its brown decayed remains, stretched upon the forest floor, may form the favorite spot for germination of seeds which will replace it. In the hemlock forests of the Lake States and in the Engelmann spruce forests of the high Rockies, it is not uncommon to find three or four old giants growing in a dead straight line, just where their seedling stages started life upon some vast old rotten log of their own kind.

In the same way as we have glimpsed the interactions which compose the forest we might view the prairie or the steppe. Once its grasses and herbs become established, they dominate and control the pattern. Into this picture other organisms must fit, or be eliminated. Grazing animals feed upon the tips of the plants, without serious injury to the protected growing points. Burrowing animals

GRANDFATHER TREE LINES UP THE YOUNGSTERS

seek shelter and obtain nourishment from the characteristically deep, black, rich soil. Legumes and soil bacteria fix nitrogen in abundance, further promoting the heavy growth.

Birds nest upon the ground in cushions of dead grass, feeding upon seeds and insects in their seasons. Carnivores and rodents range widely, swiftly, and deftly. Above float hawks and owls, looking for telltale movements in the sea of leaves and stems below. And vultures are at hand, to follow the slowing course of the infirm, and to spot the dead.

Thus does the community develop by virtue of the interactions of living organisms and become what it is—the highest expression of stabilized relationship among them.

Chapter 12

THE PATTERN OF LIFE

MORE than one traveler of the ancient world has had to await the day of modern scientific exploration to bring his vindication from the charge of plain and fancy lying. Even within the past century a noted explorer was maligned for saying that he had seen an animal now taken as a matter of course by every schoolchild—the gorilla. And the first reports of the prairies—those endless, treeless, grasslands of the midcontinent—were, to say the least, an occasion of amazement.

Curiosity to witness new kinds of plant and animal life, no less than the hope of economic gain, has been a powerful stimulus to travel and adventure. Today the variation in plant and animal life from place to place is accepted as a rule. With a few strokes the artist can symbolize the tropics by a palm tree, Africa by a lion or zebra, Arizona by a tree cactus, the arctic by a white bear, the anarctic by a penguin. This roll call could be extended indefinitely; but the question is, does it represent anything more than a list, the expression of blind chance? Is there any meaning or interpretation to be sought back of what, to the layman, certainly has all of the features of a crazy quilt? To answer that question is our present task.

To be quite honest, the pattern of life continues, unceasingly, to intrigue the imagination of men of science quite as the puzzles of master chess or bridge whist enslave the devotees of those most complicated games. And while we are behind the scenes with the scientist, let us go further and admit that, until the eighteenth century, the Doctrine of Special Creation, which states that things are as they are and where they are because they had been made so—this doctrine now regarded as so naïve and insufficient—was about the only one that would fit the facts as they were at that time understood.

For not until the eighteenth century brought us that bright, eager, busy little Swede, Dr. Carolus Linnaeus, did we have any proper technique for separating and labeling the different forms of life on a large scale. And not until life had been catalogued in some fashion was it possible to consider the patterns formed by living things upon the surface of the earth.

The boundless vitality and sure intuition of Linnaeus crystallized certain methods that had been slowly developing, and made of them a working mechanism for the classification of living things. Time and the growth of

FATHER LINNÉ CHRISTENS THE PLANTS

knowledge have given us a more flexible system than his. Yet rigid as it was, it served its purpose well, arranging the treasures of the living world so that all might see and understand them. And upon his method of naming them we have not been able to improve.

By a judgment that was quite as much artistic as scientific, Linnaeus recognized that living organisms fall into groups. These groups, like the subdivisions of a great field army, are of successively smaller size. Those most directly useful are the genus and the species. All oaks, for example, fall into the genus *Quercus,* and the various species or kinds of oak are designated by adding the proper word. Thus the white oak is *Quercus alba,* the red oak *Quercus rubra.* The lion is *Felis leo,* the domestic cat *Felis domestica.* In such manner could a name of two words—a binomial—be substituted for the ponderous string of words whereby scientists had been wont to label each species, for all the world like a breathless child describing some new wonder.

Yet it must not be forgotten that genus and species are human inventions, made in an effort to fit the facts. They are not, as many worthy people have supposed, a matter of divine imperative. Lion and cat are more unlike to most of us than horse fly and house fly; yet the judgment of specialists has put the first two in the genus *Felis,* the other two in separate genera, *Tabanus* and *Musca,* respectively, if you please. Be that as it may, with this simple device of cataloguing life, the task became largely a matter of industry and care in collecting, describing, and labeling. The resulting stimulus to biological enterprise was enormous. We now recognize over 300,000 species of plants and more than twice as many animals. The insects alone number more than all of the known kinds of plants. This progress has been made, it should be observed, in the face of the fact that there exists no generally accepted definition of the word "species." And modern studies in the breeding of plants and animals are making it very clear that the idea of a species is largely an instrument of convenience. If this all seems like a pretty ragged road to truth, ask some physicist to define for you the words "electricity" and "light."

Two other developments, beginning in the eighteenth but ripening in the nineteenth century, have been important helps in solving the jig-saw puzzle of life's pattern in space. One was the conception of geological time and process, the other the closely interwoven idea of organic evolution. The men who dared to say boldly that fossils were really the record of what had once been alive deserve to rank for courage with those hardy souls who flew the first airplanes in the early years of the twentieth century, less than a hundred and fifty years later.

As soon as earth-science ceased to be a static record of unchanging seas and lands, and as soon as the veil of past time was torn away, extending the mental view back infinitely far, the story of life itself could no longer remain static. If seas and lands had shifted, climate and soil must have changed as a consequence; and so must life too have been altered.

To the gifted Frenchman Lamarck, who observed the remarkable fitness of life to environment, the changes of the living past had been something more than a shuffle of forms already in existence; rather they implied the continuous unfolding of new forms. It remained for Darwin to make this idea generally reasonable, to establish it in fact as the keystone of modern biological theory. Thus today we view both environment and organism in the perspective of immense time, as expressions of material and energy, mutually inseparable elements in a continuously creative process.

Needless to say, classification of plants and animals, which began as a mere mechanical means to an end, has been caught up into this conception of eternal change. No longer does the classifier rest content with a convenient pigeonholing of the various forms of life; instead they are now viewed as a related, genetic, whole, and the object of the game is to classify in such a way that a true, or reasonably true, picture of these relationships is presented. Thus at every step do we find that the problem of life's pattern becomes a living, dynamic thing itself. And as Leibnitz and Newton were obliged to devise the beautiful mathematical methods of the calculus to deal with the problems of continuous change which they encountered in their studies of the heavens, so the biologist finds himself unable to use the old, simple, static formulae in the present instance. The new challenge is to deal with a moving equilibrium in terms of time, space, geological force, and evolutionary history—no small order! This being true, how do we go to work?

We first identify, name and list the forms of life, verifying them at every turn by permanent specimens. We then find, in the case of each, its range, or the area in which it occurs naturally. We then, and this is of especial importance, compare the distribution of species in the same genus, of the various genera in a family, in short of all conceivable degrees of relationship. If by rare good fortune we know of the existence of the same species or of close relatives, in fossil form, we note the fossil distribution, both as to time and place. If we find unexpected gaps, or coincidences, we consult the geologist, hoping that he may be able to tell us of changes in land form or climate which may explain the special case. And, not too seldom, we lay upon his doorstep certain problems like the existence of monkeys in Africa and South America, or of marsupials in South America and Australia, which he has a good deal of trouble in solving. He may even in his turn, as Wegener has done, go to the length of stating that continents which were once joined have broken apart and gone drifting about on the earth's surface, like meringue on a cold lemon pie.

One of the surest things about animal and plant geography is that most cases turn out to be special cases. As a general rule the species of a genus have ranges which overlap. The range of the genus is generally more or less continuous, and we explain that by assuming that all of its species have had an ancestor in common. Thus far the task is about like that of explaining the abundance of Brewsters or Aldens in New England. But should we find a lineal descendant of old Elder Brewster ruling a hacienda in Spanish America, explanations to, or perhaps from, the family gods would be in order. We should like to know how and when he was isolated from his northern kinsmen.

Plant and animal geography abound in instances very like that. Water and mountain ranges make very effective barriers. So do tongues of ice, and for many plants, areas of desert whether due to climate, salt, lava, or animal disturbance. It can happen, often enough in the course of geological history, that the continuous range of related species in a genus is broken by the intervention of such natural barriers. Species may be cut off by the making of peninsulas into islands; or by the shifting of ocean currents; or by the penetration of moving ice or the elevation of mountains. And being cut off, these forms may become more and more distinctive as time goes on; although to tell the truth and add to our confusion, they sometimes remain surprisingly constant.

In a very general way, and allowing for exceptions, it should follow that the living plants and animals on lands which have been separated in more recent times should be more closely related than those on lands which have been apart for longer times. And in the main, this is true. Old Australia, isolated since the Tertiary, amazed the world with its living museum of kangaroos, eucalyptus trees and black swans when hardy navigators had restored the bridge between it and the rest of the world with their ships. So far as anyone can tell only the dark races, the dogs, and some small rodents had gone in in the meantime. And interesting to record, on the new bridge of modern ships, the white man swarmed in with his wheat, sheep, and cattle, as well as some less desirable immigrants, to wit, the cactus and the rabbit. Parenthetically, but not with complete irrelevance, it may be remarked that the European aristocrat has private opinions as to the desirability of much of the original white stock which settled Australia. Certainly the Australian accent sounds as incongruous in a London drawing room as does the American language. But the Australians are developing a culture pattern which hangs together. And the crimes which landed some of their progenitors on prison ships might impress a modern jury less unfavorably than some of the tactics which in the past have been grounds for patents of nobility.

However, to resume the larger perspective, continents which are still in contact, as Europe and Asia, or practically so, as Asia and Africa, contain a high percentage of closely related forms. North America, which has at no very distant time, geologically speaking, been connected with Europe and Asia, shares biological characteristics with each. Many genera of flowering plants which occur in China are also found represented by species in eastern North America, while the genera of Northern Europe—apart from those brought to America by the white man—are by no means strange to the American botanist, although the species have changed somewhat with the passing of time. Species of bluebell and twinflower are quite alike, but the paper birch of Canada, for instance, is a different species from the white birch of Norway.

In the case of Africa, which itself abounds in barriers, there is a degree of biological kinship to South America, but it is certainly in the remote past. Monkeys, for example, are found on both continents, although there is considerable difference. The result of such facts is a challenge to the scientific imagination which brings serious support to two seemingly fantastic theories. One, the classic legend of the lost Atlantis—a submerged continent now supposedly at the bottom of the south Atlantic; the other already mentioned—the theory of continental drift, based upon the remarkably parallel curvature of the African and South American seacoasts on opposite sides of the Atlantic. This latter theory assumes a break which began some 25,000,000 years ago and has since been widening as these two land masses and their northern extensions have drifted apart. Whatever the final merits of these two

theories, both are significant of the challenge of biology to geological science, and a living example of the way in which these two dynamic subjects have proceeded hand-in-hand from their inception.

Ever since Darwin, and Wallace, too, observed the variation among kindred forms on neighboring islands, and adding other evidence to it, proposed the theory of natural selection, the facts of organic distribution have intrigued the thoughtful. Here surely is a setting for hypothesis on the grand scale. Some years ago the Englishman Willis brought forth the Age and Area Hypothesis. Since species originate from a single ancestor and multiply thereafter, and spread from the point of origin, why is not the area occupied by a species an index of its age? The longer it has existed, the more widespread it should be. Surely a noble and seemingly a reasonable approach, this. Yet it has proved to have one grave and fatal defect—there are too many exceptions.

All honor is due Mr. Willis for having proposed something so definite and tangible that it could be used as a shining target. Such a service is as rare and useful to science as the methodical, diligent, dispassionate observations which may later riddle the target to shreds. It so happens that many species which are endemic, that is, restricted to small areas, are very ancient. The redwoods of California, the Ginkgo of China, are relics of species which are much older than our flowering plants and which at one time were very widespread. Not able to compete with more specialized, newer forms, they have been sealed up in their present limited retreats. The behavior of such endemics is but one of many specific cases which make it impossible to utilize the beautiful simplicity of the Age and Area Hypothesis. Instead, we are forced, as has already been suggested, to recognize that each case of distribution must be investigated separately on its own merits. And we have the heartening, if by no means ingenious, assurance that where this is done by men of understanding, most cases can be explained in terms of evolutionary and geological change.

We have thus far said little about climate, except to include it as one of a number of possible types of barrier. We are of course accustomed to think of plants and animals as sure indicators of climate—we need only recall our polar bears and palm trees of a preceding page. However, we are confronted with the stubborn fact that climates may be very similar, as are the tropics of Southern Asia and South America, and yet have very different inhabitants. Organisms may indicate climate, in other words, but climate is no guarantee of the presence of definite organisms. The reason for this, in view of what has been said, should be fairly clear. But the question still remains: do similar climates, working on different evolutionary material, express their similarity in any way in terms of whatever living organisms are at hand?

To answer this question we must explore another realm of the geography of life. Thus far we have been dealing with floristics and faunistics—the kinds of plants and animals in relation to distribution. We must now give our attention to the way they fit the environment and the kinds of communities which they compose. Thus we shall find the key to our question.

So far as vegetation is concerned—and it is the clothing of the landscape—there is a striking similarity between the treeless tundras of Northern Canada and Siberia, the untimbered region south of the great beech forests near the southern tip of South America, and the treeless space between timberline and snow in the higher Rockies. Mosses and lichens are interspersed with severely dwarfed shrubs and cushiony perennial herbs with heavy storage roots and very short, much-branched shoots. We call such a landscape arctic-alpine—a term that applies on every continent which may afford the proper conditions. To the biologist it is a clear unequivocal label for such conditions. Yet if we take a census of the species which are found in arctic-alpine situations the world over, we find them, despite a few which range widely, of the most diverse relationship. In other words the *composition* of the arctic-alpine flora varies from place to place. Its *structure,* however, is highly uniform, despite this variety of living material. And this fact, when extended to the various climates of earth and their characteristic vegetation, becomes the expression of a principle. Climate controls the structure of living communities, regardless of the composition of those communities in terms of evolutionary material, that is, the kinds of living species which are found in them.

What we have said of tundra applies with equal force to the other great types of vegetation: forest, grassland, scrub, and desert. The deserts of the old and new world, for example, have been invaded by very different families of plants. The familiar mucilage-filled cacti of the American southwest are not at all closely related to the milky-juiced euphorbias of Asia and Africa. Yet both have developed a similar fleshy, succulent habit. Both are compact in their structure, heavily armed with spines, thickly coated with wax, and with a minimum of exposed leaf surface, if any. Accompanying them are hordes of tiny annual plants which spring into activity almost overnight with the coming of occasional desert rains, blooming, fruiting and vanishing with the disappearance of the brief moisture supply. Or if by chance the moisture continues for a time, these tiny plants are equipped to branch and keep on fruiting so long as the favorable conditions endure. On both hemispheres the larger perennial desert plants are characteristically widespread with shallow roots for the quick absorption of rain when it does come, and facilities for water storage, often for long periods, between rains.

Again the chaparral or the California scrub, whose occasional burning permits the rain to descend in great floods from the singed and naked hillsides above the coastal cities, is the structural counterpart of the harsh gray shrub of the Mediterranean coast in Spain, known locally as the garrigue. The success of the old Spanish padres in transferring their culture to the California coast, and the patness of their imported Moorish architecture, influenced of course by Indian craftsmanship, is much more than a coincidence. Both climates, and the vegetation which expresses them, are essentially similar. Yet the species of the chaparral and of the garrigue, even the genera and families, are entirely distinct.

We have spoken of the vegetation because it most quickly and strikingly expresses the character of the climate. But it must not be forgotten that the animals which, directly or indirectly, are supported by it are a necessary part of the community. And when we consider them we find that they too show the relation of climate to community structure. Barring Australia, which was segregated before the origin of placental mammals, we find that the great natural steppes or grasslands of every continental interior have been, at the advent of man, occupied by vast herds of hoofed grazing animals. Zebras, antelope, horses, old and new world bison—the roster changes from continent to continent, but not the essential life habit. And upon their flanks may hover coyotes, wolves, hyenas, and primitive human hunters with red, white, black or yellow skins, according to the location. The kind varies, the pattern is remarkably repeated. It need scarcely be said that what is true of the great continental grasslands is equally true for the animals of the other great vegetation types, e.g., the forest.

In the preceding chapter we looked closely into the organization and development of natural communities, as we looked earlier into the controlling environmental factors of soil and climate. Yet we would be remiss if we did not here ask again: is there not some simple, stately, formula which will, at a stroke, introduce order into this diversity and resolve the confusion? Is there not some golden principle, such as temperature, moisture, sunshine, or soil chemistry which holds the key to the pattern of communities?

Here again, as in the beautifully conceived Age and Area Hypothesis, we must confess that truth is only to be reached through the long road. To an extent it is true that the pattern of forest, grassland, scrub, and desert is determined by the available moisture, not necessarily the amount of rainfall. And to an extent it is true that within a continuous area the kinds of organisms present are influenced by temperature. But the fact remains that each great type of community has numerous manifestations, and that in a particular case the responsible environmental factor or factors cannot be known without prolonged analysis and study. Climate is the broad determinant, to be sure, but its effects are not the same on opposite sides of the same hill, or on sand and clay soils in the same farm. The temperature may be so low, on an average, that the question of water supply, important though it be, becomes a minor matter. The air may be so humid and the temperature so equable that slight differences in soil chemistry are more important than water or heat. Each community must be studied separately, and widely compared with others, to know the physical cause of its being. Even then, as we shall see later, we must be prepared to reckon with obscure events in its past history whose influence may persist to the present.

Thus far we have said nothing of man—lord of creation, yet a living organism like the pine and yucca, an animal like the bison and cougar, yet withal a creature apart. To the puzzle of his racial pattern upon the earth and the fabric of his cultures, expressed in communities, there is no glib answer. Slowly and painfully, with infinite toil the record of his early beginnings and his movements is being uncovered, but all that we know today represents the merest fragment of the grand epic of his struggles, his failures, and the joys of his accomplishment.

The world's human population today is assigned to a single species *Homo sapiens,* and this classification has been extended back to include Cro-Magnon man. But his predecessors, found to date in Europe, Asia and Africa are considered as belonging to other species of the genus *Homo.* Still earlier remains are assigned to other genera.

How far this might have stood up under the arbitrary tests which are used as standards in classifying other living animals, we can never know. Was it possible for the earlier forms to cross with the present species and produce fertile offspring? Did present man arise from these earlier known forms, or by their side, later to invade their territory and either absorb or replace their culture with his own? For one thing is certain: human culture existed before the record of *Homo Sapiens* begins.

The latter as he exists today is divided into races, which in the main are clearly enough established. Yet all of these living races can interbreed, and each shows the widest range of intergradation. Naturalists today are rightly skeptical of the claims of any considerable racial group to ancestral purity, and the scale of comparative ability among races depends largely upon the culture pattern of the man who makes the scale. An American Indian of the old school could give an intelligence test that might embarrass most of us. Language, of course, has been of service in tracing cultural migrations, to some extent those of races. But it can be overworked, for language is the easiest of disguises to adopt, if one begins young enough. Skeletal characteristics, notably of the skull, while showing a wide range of variation within a single race, are of considerable use in tracing past waves of migration.

Certainly the great land mass of the Old World contains at its margins in Africa, Western Europe, and the Indian Peninsula a layered record of successive cultural and physical types. This has led many workers to infer that somewhere in the center of this great tri-continental area there was a genetic and cultural reservoir from which wave after wave of migration rippled forth to absorb or destroy, amalgamate with, or perhaps yield to, those which had gone before.

And by miracles of nautical skill, these waves of men, instead of recoiling when they reached the shores of the Indian and Pacific oceans, broke into lesser waves which moved on in boats and peopled the islands of the sea. Some of these islands are thousands of miles from the mainland, and hundreds of miles from any way station, yet they were reached before seas were charted or compasses thought of.

As to the Americas, it is a matter of professorial routine today to toss into the wastebasket any suggestion that they could have been a really early cradle of mankind. The same may be said for the more insistent notion that African or Mediterranean man made his early way across the Atlantic, afloat or afoot on a lost continent. Respecting the doughty navigators of the Pacific the professors are perhaps a shade less arbitrary. But all agree that across the Bering, on ice, land, or boat—perhaps all three—there dribbled for a very long time Asiatics with wives, dogs, fire, and probably the art of making stone tools. And recent discoveries are tending to show that this began to happen much earlier than the three or four thousand years which used to be conservatively granted to the American Indian to accomplish his really wonderful works. Caves at the Strait of Magellan show human works associated with native horses and other extinct animals of an ancient world.

While the Mongolian-American probably spread fanwise from his port of entry, there is reason to believe, from the extremely large number of language stocks and minor culture patterns along the Pacific coast as compared with the rest of North America, that our western coast formed his main line of movement into the New World, down into Central and South America. Presently, from Mexico to Peru, he achieved civilization—a civilization which the honest old Spanish captain, Bernal Diaz, admitted to be a peerless thing in many respects. Meanwhile more primitive cultures were being established elsewhere, until from the cultural centers of high and low degree there began to radiate new waves, sometimes to persist, often to recoil back against the parent center. The result, by the fifteenth century was a condition of balanced equilibrium as taut as a harp string. And thus, when the European culture, in the last great wave of its age-long western push, stepped upon the eastern shore of the New World, the string snapped, and profound displacements occurred far into the interior, decades ahead of our arrival there.

In Oklahoma, where this is being written, one can witness today the final welding of this spherical web of humanity which has girdled the earth in its broad band. Here, more truly than in Singapore or Malay has West met East, for here the two great waves of racial and cultural migration have come to an end. There are no more frontiers—no more wilderness, unless it be the wilderness of man's own creation.

Here one may find error, imperfection, and injustice—the inevitable confusion when alien patterns merge, disintegrate, and struggle to reform. But the worshiper at the shrine of Nordic, or any other racial superiority will have to seal his eyes to the beauty of the women, and avoid inquiry into the ancestry of many of the leaders in arts and affairs, unless he stands ready to part with his cherished illusions.

Chapter 13

SUCCESSION, THE GROWTH OF COMMUNITIES

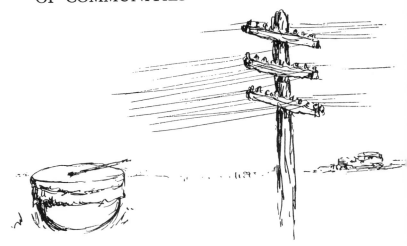

HUMAN history, properly understood, is something much more important than a chronicle of egocentric and bullheaded trouble makers. The terse reports of Julius Caesar will always interest the professional soldier; but their really priceless feature lies in their casual glimpses of barbarian communities which have since become the urbane and metropolitan centers of western Europe. Each of these centers, as Rome had done long before, has changed from an assemblage of fairly equal and independent individuals to a highly specialized, highly stratified organization.

In the United States a corresponding development of communities has taken place rapidly enough to be a matter of familiar fireside tradition. Modern technical invention has so telescoped the process of succession that almost any reminiscent grandfather is likely to have a good deal of genuine history under his hat. In fact the momentum of change has been so great and so little hampered by the inertia of a long past that conservative Europeans are apt to speak querulously of "Americanization" when their own communities undergo any great technical readjustment.

But the rate of change and the details of its accomplishment should not blind us to the fact that there are laws of community development which apply rather generally to all living things. And before we examine the special case presented by human society, it will be well to consider the general problem of communities in nature. These communities include both plants and animals, yet their character is tinctured, if not controlled, by the dominant vegetation. Forest, grassland, scrub, and desert with their various subdivisions form the essential matrix of community life in nature.

Each in its turn consists of a great diversity of smaller communities. The North American forest can be subdivided into needle-leaved and broad-leaved—the softwood and hardwood of the forester, respectively. Taking the former, the "evergreen" woods of common speech, we might begin with the Canadian spruce-fir forest and circle about the continent, boxing the compass with different phases at every turn. For in the lake states white, jack, and red pines are characteristic; in the Northeast spruce

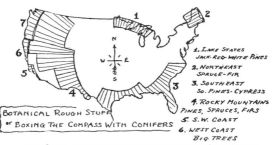

BOTANICAL ROUGH STUFF or Boxing the Compass with Conifers

1. LAKE STATES JACK-RED-WHITE PINES
2. NORTHEAST SPRUCE-FIR
3. SOUTHEAST SO. PINES-CYPRESS
4. ROCKY MOUNTAINS PINES, SPRUCES, FIRS
5. S.W. COAST
6. WEST COAST BIG TREES
AND 7. THE MOST MAGNIFICENT N.W. PACIFIC COAST FOREST.

and fir; along the New Jersey coastal plains scrub pine and pitch pine. Farther south these give way to the southern yellow pine, the long-leaf pine with its golden purple trunk and its rich store of turpentine, and the feathery, leaf-shedding cypress which transforms the swamps of the southeastern and the Gulf states into stately green cathedrals. Then leaping from western Texas into the southern Rockies we would encounter junipers, pines, firs, and spruces, neatly filtered and arranged by the selective forces of altitude and exposure. And next, on the western coast, we would pick up in southern California a gnarled and straggling group of pine and cypress, then scattered stands of those living fossils, the redwoods, as we moved north, finally completing our circle in that superb crescendo, the

[194]

northwestern evergreen forest. Here, stretching from Oregon to Alaska, are the magnificent reaches of Douglas fir, Sitka spruce, and western hemlock.

Likewise, we could follow the hardwood forest west from the Atlantic, rich with beech, birch, maple, and tulipwood, through the oak and hickory of Illinois to the black, twisted, scrub-oak forests of Oklahoma and on into the knee-high shinneries of the Texas panhandle. Or if we chose the prairie, and moved north from Texas into Dakota, we would pass from lush fields of bluestem, and Indian grass, embroidered like blankets with paint-brush and blue-bonnet, and move gradually into stiff wire grass, porcupine grass, and wheat grass, dotted with buffalo bean and loco weed. It would be like moving one's hand from the silky coat of a Cocker Spaniel to stroke the back of an Airedale.

In the high plains with their short grass, the woody scrub of the chaparral, or the great deserts of the Southwest, we would find a similar diversity of detail within the general uniformity of aspect. And the same principle applies on every continent in the world.

Moreover, the boundaries among the major communities we have named are seldom unbroken. The eastern hardwood forest, with its characteristic mammals, birds, and insects, sends its runners far west into the prairie states, along the flood plains of their rivers, and even upon the sandy uplands of the southern midcontinent. Conversely, the prairie extends eastward in diminishing patches through Ohio and Indiana, Kentucky, and Alabama, even reaching the seacoast in tiny areas of Long Island and Florida. What hope of order is there in such a madman's design?

Let us turn for a moment from the communities to the plants which compose them. There are, for example, some hundreds of species of trees in North America, practically all of them peculiar to the Western Hemisphere. Some of them range widely, the bur oak for instance being found from the Appalachians westward until it meets the advance guard of the Rocky Mountain forest in the stream valleys of Nebraska. Other trees, like the Monterey cypress of the California coast or the Osage orange of the near Southwest, have a very restricted natural home.

Inside the limits where a tree may be found there are many situations not favorable to it, others which might be favorable enough if the soil were suitable, or if other vegetation were not on the ground. Sometimes the range is so great that the rôle a tree plays in one place will differ greatly from that which it plays in another. The bur oak which we have mentioned prefers the well drained hilltops in the humid East, but requires the protection of sheltered, moist and fertile valleys in the dryer West.

Naturally what we have said of trees applies with equal truth to other kinds of plants, and to animals in spite of their propensity to roam. The communities of a continent must be made up of the living stuff which is at hand. The first and broadest selection is due to climate, some organisms tolerating a wide variation, some enduring only very narrow extremes.

Within an area of suitable climate the surface of the

earth is varied, dissected by valleys, elevated into hills. These differences affect the impact of climate by their shelter or exposure. Light and wind, for example, are much more severe in their effects upon the hilltop than in the valley below. Topography also influences greatly the distribution of soil-moisture and other materials essential to life. The soil itself thus varies with topography. Valley farms are the first choice, hills the last resort, of the shrewd pioneer farmer.

Thus is it inevitable that communities of plants and animals must be intricately divided and interwoven. Diversity of condition is the rule, diversity of living communities the consequence. This would be true if the landscape and climate remained static in their present condition. They are, however, dynamic and changing. Thus our problem of interpretation is one of understanding a moving, rather than a stable, equilibrium.

And finally, communities do not spring into being fully established, whatever the condition they have to meet. They must grow and develop into an adjustment in any particular place, however permanent or temporary the conditions there may be. This process of adjustment is known as biological succession, and it applies to Naples or St. Louis quite as it does to the dunes of Lake Michigan. Forever it is taking place against a changing physical background. The restlessness of Nature is like that of a living creature, always tending towards what it never attains—balance, adjustment, repose.

But this principle we shall understand better after a little spade-work.

In our search for order, let us borrow a leaf from that tidy chamberlain of all the sciences, the mathematician, who never goes to work without making an initial assumption. Let us assume that we have an area—a sort of biological Graustark—whose climate is uniform throughout. Let us assume another thing which never happens, that the entire surface is uniform, bare, and smooth, but that plants and animals are in the offing, ready to move in and take their chances. What will happen?

Assuredly, animals will move in and have a look, excepting those kinds which never leave the shelter of vegetation. They will, even on brief excursions, leave their droppings, and if followed by their enemies, perhaps a portion of their remains. Birds will fly across, and insects, always fecund and often requiring little food in the adult, winged condition, most assuredly will swarm in to mate and die, perhaps as prey to birds or other animals. Some insects such as the beetles, will find congenial home and food in the organic material left by larger animals.

Seeds will enter on the wind, or be carried in by the visiting animals, but their entrance is as yet in vain, for they cannot root unless it be in some moldering pile of animal material. They are much more likely to be eaten or to be rotted by the fungi whose spores have come in with them. Along with these seeds and spores, however, come the fragments of lichens, whose curious, scaly growth needs no root—only a rock surface on which to cling. These, in all likelihood, will be the first anchored life.

Meanwhile the sun shines and the rain falls. Day and night, winter and summer, alternate. No rock, not even eternal granite, can endure these simple changes without effect. Like the stolid Chinaman under the gentle drip of the water torture, the particles of rock yield and slowly disintegrate. Because of the scattered lichens and heaps of animal material, the uniformity of its exposure is broken. As the rain falls and flows it washes gullies in the loosened bare rock, but soaks into the organic material as into a blotter. Inevitably these gullies form a system and drainage is established. Here and there the wash brings together an accumulation of rock particles and of organic material. Thus soil is born.

Now seeds can sprout and take root. Many do, but only those kinds which endure the strong sunlight survive. Even these, however, provide some shade and considerable food for the appropriate animals. Great cracks appear and grow in the rocks, and the newborn soil which accumulates in them allows the pioneer plants to root deeply. As these roots penetrate and die, burrowing insects and worms, fungi and bacteria follow downward to feast. Attracted by the growing permanent population of small animals, larger ones come in their turn, for food is the great mover of animal populations.

Thus is established a pioneer community, able to survive in spite of exposure to the elements and the absence of anything but the simple beginnings of a soil. Meanwhile the diversity caused by cutting streams and remaining upland grows greater. The pioneers are continually having to make a new start as the upland is washed bare, but within the valleys are shelter from wind and sun, and in

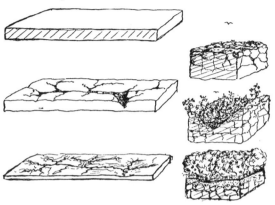

places deeper soil material. Here the shade-lovers and their retinue can thrive. Even in places on the upland, level, protected spots where the soil can grow undisturbed, the pioneers themselves produce so much shade and competition that their own offspring cannot live, but give way slowly to shade-loving plants.

We started by assuming uniformity, but we cannot maintain it in the face of dynamic events. The situation is getting out of hand, so far as simplicity is concerned. Change rules—where does it stop? The answer is, it never stops. Where, then is our promised order?

To find it we shall have to take apart our wilderness, which started as even as a billiard table. Let us first examine the valleys and hills, work of water and its sister climatic factors. Steadily this work goes on. Barring interruption, which is our privilege, the narrow, deep, gorges become broad and the hills become gently rounded. In the end we have something which is as near a plane surface as nature can produce—a peneplain, it is called. Drainage of the peneplain is thorough, but not violent, for the slight slope, absence of relief and complete system of streams and tributaries insures that it be so. Water does not fall to rush off at once, but makes its way leisurely. And because the strong contrasts between hill and valley are destroyed, climate plays more uniformly upon the surfaces everywhere. We have neither the deep shelter of the valleys, nor the severe exposure of the hilltops and of the original surface.

Let us consider the surface itself. The rock has continued to weather into fragments, and slowly these fragments themselves have changed, physically and chemically. In them generations of plants and animals have done their work, until gradually a mellow, fertile, soil has formed. Unequally distributed at first, the gradual leveling of the area has lessened this discrepancy of soil pattern. The ceaseless activity of minute plants and animals has played its part. Roots have drawn minerals to the top, where they are returned as fallen leaves, and otherwise. Surface animals have transported substances back and forth. Burrowing creatures have likewise carried stuff up, down, and around until the whole top horizon of the soil has been worked as thoroughly as a haus-frau mixes dry suet and flour to make a pudding. As a result the surface has become an organized soil and has acquired many properties which the old smooth parent rock did not have. Easily permeable to roots, livable for them and much else that is alive, it also holds water like a sponge. Now the streams flow clear, for they are fed the rainfall from this filtering reservoir instead of from the barren, muddy hillsides. Here, too, within the surface layer, as in the topography of our area, sharp contrasts have been softened and extreme differences eliminated. So far as conditions can be uniform at the surface, they have become so.

Finally, let us see what has happened to the communities of plants and animals with which we started. When we laid them aside they were changing, with no repose in sight. How have they fared?

Perhaps the most surprising development is the fact that the tough, hardy, and venturesome kinds have been replaced by others which are less so. So accustomed are we to thinking that organisms which can stand any kind of rough-and-tumble must always, in competition, have their way that this requires explanation. It comes in terms of efficiency. Under the improved conditions of soil, water distribution, shade, and needed fellow organisms, the latecomers thrive better than the pioneers. Truly, they are more sensitive and more exacting in their requirements. But once these requirements are met, they are more effective, able to grow taller, root more deeply, live longer, reproduce in conditions of their own creation, and utilize or at least tolerate, more of the other accompanying organisms.

This is to say that the plants and animals have shown succession, beginning with pioneer stages and slowly progressing, until a complex has developed which is in equilibrium with soil, climate, and fellows. This essentially stable complex or community is known as the climax. With a little experience, anyone may observe the stages of its development in an abandoned roadway, meadow, or cut-over forest. One sees first the annual sunflowers, foxtail or other weeds, then the gradual entrance of perennial plants, perhaps the ragweed or gumweed, or it may be sumac or hazelbrush. At length, after the first pioneers are out of the way, one may see some venturesome native prairie plants or young forest trees here and there, the first indications that stability is once more approaching. These are only the most obvious changes, for the kinds of insects, birds, and mammals, as well as fungi and bacteria undergo corresponding succession.

Nature comes as near giving a guarantee of permanence to a species as she ever does when she confers upon it the privilege of membership in a climax community. Fortunate are the kinds which are so constituted that they do not poison, or otherwise make unfit, the habitat for their own offspring. On the other hand, the increasing areas of human disturbance tend to play into the hands of pioneer forms; and weeds are as constant an index of civilization as tin cans and profanity.

What the climax will be on our imagined land depends

upon the kind of climate with which we choose to endow it. As a rough clue to what this means, if we had imperially decreed a humid climate, the course of events would in all likelihood have led to a forest. A subhumid climate would have brought lush grassland, or perhaps savanna—that park-like mixture of forest and meadow. A semi-arid climate would only permit the development of short grass or scrub, and aridity of course would have meant desert.

Such is our order—an order of trends and tendencies. Everywhere topography, the floor on which life is deployed, strives to "exalt the valleys and make low the hills" as it moves and tends toward maturity. Everywhere, likewise, soil tends toward a condition of maximum productiveness. And finally, life itself tends towards a combination of ultimate richness and balance.

Here, it seems, if we have no description, we have at least the tools for it. By wide and thorough study (for our imaginary land is but a simplified picture of many detailed glimpses which scientists have obtained), it is possible to judge how far these idealized trends have carried. They do not always march in step; far from it. But they are there as a dependable aid to understanding which works when tried.

Thus living communities are stages along the way towards an ideal which, if achieved, would be as little handicapped by imperfections of soil, climate, topography, and organic adjustment as it is possible for life to be. And so, when in the field we face a maze of communities, while the task of mapping them is perhaps as intricate as if we knew nothing about succession, yet interpretation is possible. One community, if time goes on, may develop until it resembles a second. A different one may be incapable, under the existing conditions, of progressing further, having reached a climax or equilibrium.

It must be admitted that we are simplifying more than Nature usually does. She is rather given to interruptions and reversals of her own smoothly charted course. Lands submerge, the sea bottom rises, drainage patterns shift and the cycle of erosion may be renewed. Volcanic ash or lava has often, as in the Yellowstone or Alaska, entombed all that is alive and the lengthy adventurous round begins again. Yet there is this cheerful side to it, that precisely such fresh starts have shown us how the process works.

These jugglings of the terrain are not all. Climate, too, can change and alter the course of events as we are beginning to learn. In fact, science has a good deal of knowledge on this subject which might prove useful to the business man and farmer. Most of us, however, have our hands so full trying to meet the emergencies caused by unfavorable years or cashing in on the good ones that we have no time to learn whether they are a matter of order or pure chance. While climate is a fluctuating thing, the extreme years tend to fall into groups, thus intensifying their effects. The perennial cry "We'd be all right if we had some rain" indicates that enterprise is usually planned as though the best were the normal. The years of drought are a part of the normal, and must be reckoned with as such. In 1936 following three successive years of drought, the soil of the Nebraska prairies was so dry that their composition began to shift; the advantage was with grasses usually associated with a dryer climate. But in 1937 rains restored the Wyoming range to beautiful luxuriance that had seemed impossible the year before.

Over longer periods, measured in centuries rather than decades, the climate takes bigger swings, with correspondingly bigger effects. Sometimes these result in major shifts of the boundaries between great communities. The intense dry, warm period of about three thousand years ago brought the prairies east to Ohio where some of their remnants have lingered. It also moved the timber line higher up the Snowy Range than it is today, and probably completed the melting of immense blocks of dead ice which had lingered in the mountain tops.

Even with these great shifts whose measure is long we are not done. The periods of geological history have varied in their broad climatic tendencies. The Coal Age, as we have seen, must have been a time for rank, lush greenery over much of the earth, as was later the Cretaceous. The Permian, on the other hand, and the Pleistocene from which we have so recently emerged—if indeed we have—were characterized by climates of great severity.

Naturally the effect of these major geological fluctuations is much more profound than that of changes of lesser order. Communities are shifted, often exterminated on a grand scale. And this introduces us to another, most potent factor which breaks the simplicity of our imagined realm with which we started. Each great geological period has witnessed the wholesale development of new forms of life and the concomitant extinction of others.

Thus back of all the complexity we have set forth, lies the fact that nature is continually evolving or developing new kinds of plants and animals to compete with those already established. Even if habitats were stable and succession orderly, we would have to reckon with this more or less continuous supply of new living material to participate in the process of organic succession and community formation.

Thus proceeds the eternal search of life for the thing it never finds: a perfect equilibrium. And yet, just as our friend the mathematician makes constant use of infinity and of functions which approach limits they never meet, and by so doing creates an intelligible, consistent whole, so may we. For us, ideas like that of the peneplain, the stabilized, mature soil, the climax community with its organic equilibrium—ideas which do not often materialize—may serve to fit together the diversity which is spread before us. This sort of reasoning, no less than the meticulous accumulation of facts, is a function of the human mind in science. And let it be recorded that the laborious calculations of the mathematical physicist, which today are the world's byword for profundity, are simplicity itself compared to the material with which the student of organic life must deal.

Into such a world as we have tried to picture—a welter of communities, each an expression of struggle towards stability in the face of the utmost diversity of environment —man has entered. Man, Prometheus, has dared to break this universal order. He is, for the time being, the really dominant organism. True, he lives by the corn in his fields, but he plants as he wills, not as nature decrees. Whether he can sustain this revolution which he has initiated is a question for the future. His power of reflective intelligence may decide the issue, but not necessarily in his favor. A two-edged sword can cut both ways.

PILGRIM'S PROGRESS

THE PATTERN
OF HUMAN CULTURES

Chapter 14

MAN, THE HUMAN ANIMAL

"WHERE self the wavering balance shakes, 'tis rarely right adjusted." In other words, the most difficult thing for a man to understand is himself. We can capitalize the word Man and omit the article without changing the truth of this statement. At once the most difficult and most important task which faces mankind is to know itself better—far better than it does today.

This, it may be observed, is no small order. The physician must go through years of stiff apprenticeship to learn something about the human body in health and sickness. He must carve up dead bodies, studying their construction and chemistry. He must observe the workings of the living body in laboratory and clinic. And on top of that he must be gifted with a certain artistic insight and subconscious understanding if he is not to be a menace to his patients.

The psychologist tries to know the workings of the central nervous system of man. To a degree this is a matter of experiment and measuring. But a machine has no significance apart from its product, and the product of the brain is thought. Our best efforts cannot separate psychology from philosophy. Try as we will, an impersonal, scientific analysis of the human mind remains a convenient fiction, at times quite useful, always extremely hazardous.

The historian, the linguist, the artist, and the philosopher (so far as you can cut them out from the rest of the herd), concern themselves frankly with the products, rather than the machine which produces them. Of these products, science alone aims at predictable certainty. Action, imagination, beauty—these creations of mankind are unpredictable, endlessly fascinating. And so Science, with her workaday gropings at the things she can see and feel, has become the Cinderella of the humanist, even while she is transforming the material world about his head.

No son of the daughters of men lives to himself. The physical bond which is broken with the severance of his umbilical cord becomes supplanted by others, less material, but equally binding, more profoundly controlling. The stuff of his growing experiences is that which his group provides. He cannot think apart from it, for the words which he learns to use—the tools of his mind—are those of his group. Even the lower animals are caught in the behavior pattern of their fellows. The carrier pigeon cannot use his marvelous sense of direction until he has flown with the flock in ever widening circles and thus learned the terrain about his native cote. The lion cub acquires his craftsmanship as a hunter because of parental solicitude in providing partially stunned victims for dress rehearsal. A magpie taken early from its nest may grow quite large without having sense enough to feed itself.

But these are simple things compared to what we call distinctively human attributes. Such attributes come only by living among people, no matter how rich the capacity we inherit at birth. The experience which really works upon a man to make him what he is is that made possible by his fellows. This is the most effective part of his living environment. And the study of this living human environment is the task of the social sciences. Some of them, like economics, are concerned largely with certain mechanisms in the cultural pattern. Others are preoccupied, as is much of sociology, with immediate practical problems. Thus the work of synthesis, of creating a whole view free from distortion, is very difficult.

Back of the living human environment, moreover, and inseparable from an understanding of mankind, is the rest of the natural environment. This is as compulsive for us as it is for the beetle of the dunghill or the seedling pine of the forest. But its action is less direct, more obscure, because it must work upon us through the fabric woven by our particular culture.

It is one of the harsh surprises of life, particularly to the man of independent temperament, that we can have no conscious experience that is quite divorced from this cultural fabric. A slap on the back would seem to be just that —a slap on the back. But it is one thing at an Elk's barbecue, and a gravely different thing in a community of swaggering, cocky duellists. There is no experience, unless it strikes a man unconscious, that does not filter into us through the medium of our culture pattern.

This is but half of the picture. For while environment works upon us through the culture pattern, we in turn work upon it in terms provided by the same medium. That is to say, apart from certain insensate things we might do to the earth or a fellow being, we are governed by the fabric of which we are a part. The Indian, operating on a subsistence basis, was essentially frugal in his use of natural resources. Operating, ourselves, within a pattern of exploitation, we exploit. Frugality with us becomes a synonym for antisocial behavior. Everybody hates the tightwad, unless it be just before his will is read. And yet so fundamentally unsound a thing is waste in any department of experience, that those very concerns which prosper by encouraging extravagance in the buying public could not remain in business a month if they did not enforce the strictest economy in every detail of their own operation. As for the public thus beguiled, most of us in literal truth do not stay in business under the prevalent system.

We have digressed, yet not entirely without excuse. At any rate enough has been said to show that human culture lies very close to the heart of man's relation to environment. In consequence we shall have much to say about it in succeeding pages. But it may be well to approach this potent subject somewhat gradually.

Geologically speaking, man is a brash newcomer to the

earth. It is very flattering to ourselves to think that we are the capstone of evolutionary change—the result of countless ages of trial and error. We must be good, for we are the latest model. But just as the newest model of streamlined, dog-scraping automobile is made to run over a continent paved with ribbons of concrete and asphalt, and not over the bumps and mud of thirty years ago, we, too, are the beneficiaries of a prolonged modernization of the world around us. In a world of conifers or ferns and reptiles, without the cereals, without any of our other crop or vegetable plants, without any of our modern domestic animals whose physical needs, after all, run so nearly parallel to our own, we could be little more than a furtive, abject race of scratchers, gnawers, and eaters of the very things we try to keep out of the kettle.

Certainly there are two sides to man's place in nature. He himself is unquestionably a work of relative perfection. But so is the world into which he fits. The two have grown up side by side. And the modernized earth is far more essential to man than he is to it. Earth has calmly witnessed the passing of many very interesting kinds of life when they ceased to be fitted to her conditions. Certainly man should be somewhat cautious in dismantling an abode so admirably fitted to him, for in the very nature of things he is not the one to have the last word.

The modernization of the earth, in terms of plant and animal life, was about completed at the end of Cretaceous time. But not until the Miocene do we have primate remains which suggest proper ancestry for man. By the end of the Pliocene there are indications of definitely prehuman material, and the great Pleistocene or glacial period is the background for primitive man himself.

This last great sweep of time, which may have been anywhere from a half-million to two million years in length, was marked by several prolonged advances and retreats of the polar ice caps. At least one of these periods of retreat was marked by a lengthy interval of climate warmer than that of today; while during the advances a large part of the now habitable temperate zone was affected. The life of one primitive man, or even of many generations, could not have witnessed much change. But the race itself was obliged to make many adjustments which undoubtedly enriched its cultural capacity.

It is entirely proper that we should respect the resources of the human spirit. But it is well to remember that these resources are in large measure the result of a very simple and yet very marvelous change in animal design which has set man apart. The development of the grasping forelimb in the primates seems unquestionably to be connected with the habits of living in trees; we can pass up the baffling problem of cause and effect for the present. The possession of four limbs for walking and running is a very great insurance of stability to an animal which spends its time on the ground; it takes more than a push to upset a horse. But to swing about in a tree with freedom and still retain the power of walking when necessary an animal requires arms and paws, legs and feet.

For simple climbing it is enough that the forepaws be

good hooks, capable of serving as feet on occasion. These are among the marks of the ape, and if we had gone no further, we should still be at his level. The real event in our own checkered past which we ought to celebrate with gratitude was the origin of the thumb, and the changes in skeleton and muscles which forever released the forelimbs from the mechanical burden still imposed upon the human foot. True, man has no monopoly on grasping or-

THIS FELLOW OUGHT TO WEAR THE DIAMONDS BUT HE CAN'T BE BOTHERED!

gans—witness the crayfish and lobster, or the eagle. But man is the only animal with a highly developed central nervous system whose grasping extremities are freed of other duties, to serve for manipulation solely. So far as his sensory, digestive, and reproductive equipment are concerned, there is no reason to think that these are significantly different from those of the prehuman level. But once man gained the power to handle, and to do, he became man. And we can be quite certain that these activities stimulated the nervous system into development and enrichment—that is a commonplace of the modern laboratory. Action certainly registers itself in the nervous system.

It is well to remember these earthy details about ourselves.

Of the results of this miracle of the thumb we have only glimpses, as we do of all of the past. We do know that man of the Ice Age was no longer a tree dweller, for he has left signs of his activity in the caves of southern Europe, among other places. The tools of stone and bone which he shaped have survived his other handiwork. No doubt he was equally clever with skins and wood.

The first objects suspected of being his handiwork are so crude that we can never be sure whether they were the result of design or accident. There is always a great hubbub whenever anyone reports the finding of eoliths, supposedly the crude chippings of "dawn-man." But in the course of time came craftsmanship. The Folsom points of

DAWN-STONE?   FOLSOM POINTS.

northern Colorado, buried under two layers of mature soil profiles, and hence not recent by any means, are perfectly channeled on either side. And the consummate artists who made them were able to produce this long concavity by removing a single piece, possibly with a single pressure of their horn tool.

Before man left the caves of southern Europe he began to draw and paint upon their walls, to engrave upon the bones of animals the scenes about him, and even to model in the wet earth of the cave floor. Thus he, though long dead, still speaks vividly to the man of today.

By the end of glacial times the races of man had begun to differentiate, and the cultural patterns too, so that the rest of human history is a complex story of development, migration, competition. Concerning events in places other than southern Europe, we are not well enough informed; but the evidence is slowly being mustered to show that man was already widespread and diversified.

The great steps of man's cultural progress are the simplicities and commonplace of today. Yet they, and not the swinish brutalities often called history, are the key to the present. The taming of fire and the cooking of food invested the home with ritual and order. Fire comforts and warms, but it also purifies. The heat of its flame and the creosote of its smoke kill the bacteria of disease. Why should not the followers of Zoroaster have worshiped it if they pleased?

The tanning of skins, the invention of weaving rushes into baskets, and of fibre into fabric gave us the things without which modern life would be unthinkable. Our technicians of the modern world are largely concerned with doing the very same things on a grander scale.

The earliest pottery bears the imprint of basketry, as though baskets had been lined with clay and then burnt. And long after the art of pottery had been mastered, the imposed design of the basket weave persisted, as quite useless Gothic arches survive in our modern steel-framed churches.

The stories of transportation and architecture which seem to be reaching their apotheosis in airplane and skyscraper are a continuous record of using the materials at hand to solve immediate problems. Boats and wheels, blocks and boards, playthings of the growing boy, were the serious playthings of the growing race. Dugout, canoe, sailing vessel came in their turn, long anticipating the logarithmic calculations of the modern engineer. Caves, crude piles of flattened rocks, and log shelters roofed with sod were in the end supplanted by the superb masonry of Peru and the Near East. Without mortar, the Peruvians learned to shape their stones so perfectly that the blade of a penknife cannot be pushed far into the joints.

Not the least remarkable feature of all of these achievements has been the fitness, the economy of line, and the beauty—satisfying even to the captious modern—of design in tool, in weapon, utensil, means of transport, and place of abode, which is the mark of perfection and adjustment. Ugliness in the works of man there is and has been, in plenty. But invariably it represents his beginnings and his uncertainties. Achievement, whether in the bark canoe of the Iroquois, the jade chisel of the Maya, or the violin of Italy, has beauty for its hall mark.

But we must come back once more to earthly matters, this time to the ceaseless problem of filling the human belly. Man could, and did, develop remarkable cultural traits while he was still dependent upon the wild bounty of nature for his food. But like the tigers of India or the tree cactus of Arizona he had to live widely spaced to keep alive. Generally he had to keep extremely busy, and not seldom he was obliged to go hungry. Neighbors were a menace to his precarious food supply; extra mouths in his own community were a costly luxury often to be disposed of by infanticide or killing of the aged. Constant guerilla warfare, formalized into a dangerous and exciting game, was a necessity. It served the same biological purpose as the killing of surplus twigs and branches by the shade of those which survive, thus giving the whole tree some measure of symmetry and balance. There is every reason to know that the simple cultures of hunters and food gatherers maintained a realistic, working relation to environment. Otherwise they could not long survive. The preservation of a balance of nature was an essential condition of their existence. Small wonder that the Europeans found the forests, the game, and the soil profile of North America virtually unscarred! Not the least serious consequence of the Pax Britannica has been the multiplication of pacified tribes beyond the means of subsistence.

Domestication of plants and animals created revolution in the ranks of mankind. Its beginnings are lost, like other beginnings, in the mists of ancient time, but we have suggestions as to how it must have developed. Volunteer plants often come up thickly from spilled or rejected seeds today. Squash, wheat, and peach trees are familiar instances. Farm children are in the habit of taming the young of wild animals whose elders have been killed—coyotes, bears, squirrels, even crows. Often these pets revert to the wild, but evidently in the past there have been many exceptions to this rule. Even today the cat is less completely domesticated than the dog, and the Hereford is tamer than the Texas longhorn which it has so thoroughly displaced.

The most fortunate feat of domestication was that of learning to grow the cereals. In reality this was a multiple feat, for the three greatest cereals, wheat, rice, and maize, were placed under cultivation independently in widely different quarters; and with each was developed an accompanying complex of other nutritious plants.

It is generally believed that the cultivation of plants must have preceded the domestication of animals. An exception is the dog, which by his service in the chase added to, rather than drew upon, the food supply. His early association with man is attested by the marks of his teeth upon the broken bones of early refuse pits, no less than by his complete and intuitive fitness as a companion and aide in the whole range of simple domestic economy. So far as cattle and goats are concerned, there could be no advantage in taming them until surplus food was available, so that their milk might be obtained. Wild horses and hogs, too, were doubtless conquered the more easily by palatable supplements to their natural forage. It is not easy to visualize primitive food gatherers or hunters without cultivated plants, which domesticated animals need so abundantly.

Yet once domesticated, livestock could be moved about in herds on the great natural grasslands where its own wild ancestry had roamed. In this manner arose the great herdsmen cultures, which made but incidental use of agriculture. Moreover cattle, besides supplying the necessities of life to their owners, served as a valuable medium of exchange for the products, agricultural and manufactured, of more densely settled regions.

This mode of using the grassy, continental interiors had numerous important consequences. It served to defer until comparatively modern times the use of such regions—the richest potential grain farming areas in the world—for cultivation. It also fostered a restless, energetic, and eventually aggressive type of culture which could only persist in place under two conditions. Population of man and beast could not be too dense—at least ten acres to an animal is a conservative figure—and climatic conditions had to remain suitable for the production of ample pasture. That neither condition could always be favorable is amply proved both by science and the events of history.

In Europe and Asia the domestication of the horse—which had flourished in the Americas during the Tertiary, and even later—was an added factor of great importance. This animal is extremely useful in the custody of roaming herds of sheep, cattle, and goats. There is, moreover, as military men have long recognized, a powerful psychic influence arising from the association of man and horse.

Dash and verve, combined with orderliness (not necessarily cleanliness) are perhaps its most distinguishing features. The animal must be controlled, which requires self-control of his master; and the horse is of no use unless systematically cared for, which demands the same quality. Quick and frequent movement demands economy of equipment, designed and selected for efficiency. The result of this is order. And finally, the stimulating effect of speed, elevation above the ground, and the momentum of the mounted man as compared with the unmounted are obvious.

While the nomadic cultures were thus developing their character, domestication of wild life was producing significant results elsewhere. Along the lower reaches of great continental rivers, where abundant natural rainfall or the possibility of irrigation were to be found, there developed intensive agricultural civilizations. The continuous supply of fertility and usually of water from vast regions upstream kept these enterprises going, thus giving rise to the fallacy that such systems of agriculture were permanent and self-contained.

The areas suited to such intensive agriculture were highly productive, but also definitely restricted. Population tended to multiply rapidly due to the abundant production of food, but it lacked any ready opportunity for migration. Highly developed caste systems arose from the need of maintaining internal order and peace. Labor became abundant and cheap, making possible—if not making necessary—stupendous public works, such as the pyramids. The unemployment problem is no child of the twentieth century. Human life came to be regarded lightly, and for the average man held so little of joy or promise that he came to think much of a hereafter, either of blessed oblivion or happy fruition.

These teeming fertile valleys, because of the abundant leisure and labor they provided, came naturally to exhibit considerable development in the arts and crafts. The necessity for calendars and surveys led to marked progress in astronomy, mathematics, and engineering. Often considerable chemistry, of a rough and ready sort, was applied, as in dyeing, medicine, and particularly in metallurgy. Mining had been practiced in the age of stone as a means of obtaining flint, so that the problems which arose with the use of materials were more those of fabrication than of production.

In turn, copper, bronze, iron, and steel supplanted the primitive stone, often without much change in the essential design of the tools and weapons produced. In the magnificent jade technique of the Chinese the stone age art has come down unbroken. Elsewhere, however, the break seems to be complete as the newer materials utilized the old designs for practical ends.

Each improvement in the hardness of metal used literally gave to its inventors an "edge" on their rivals. The bronze of the native Britons crumpled before the steel of the Roman legions, just as in other places skilfully chipped stone weapons had failed in the presence of the keener bronze. Moreover, the invention of smelted metals, and particularly the alloy of carbon and iron which is steel, led to a widespread cutting of forests everywhere, in order to produce charcoal. Thus the destruction of timber in the Old World, with its ultimate wood famine, proceeded far faster than the demand for arable land would have required.

Naturally the art of metal-working, besides its military uses, gave rise to the production of articles for export, thus supplementing any surplus of food which might have been available for commerce. The great valley civilizations were well located for such export by water. Material for ship building was either at hand or could be brought down from upstream. Progress in maritime art thus came readily. This in its turn led to exploration and slow colonization, leading eventually to intercommunication between continents. Forgotten contacts, lost as a result of devious land migrations, were thus restored.

Commerce by land fell largely to the lot of the pastoral nomads, supplied as they were with beasts of burden. Accustomed to movement, familiar with the continental routes among which they lived, and skilled in the care of the livestock which was their chief possession and which had to be moved on the hoof, they developed the caravan as a technique for overland transport.

With the passing of time the bulkier commodities, grain, ore, and lumber, for example, came to be moved largely by water. Wealth in its more concentrated forms—slaves, jewels, spices, and costly drugs—was most profitable to the overland routes of commerce.

Meanwhile the progress of invention, while steady, was slow and casual. The ancient world was dependent upon trial and error, for it did not have the powerful, calculated method of modern science at its disposal. Thus there was ample time to absorb each innovation as it came, and make the necessary adjustments to it. There was not the technical and domestic instability of today, when revolution follows revolution within the routine of a single lifetime.

Instead of the technical changes of today, population increase, with its surplus labor and pressure upon the means of subsistence, was the dreaded foe of equilibrium and order. In the great valley civilizations we have discussed, standing armies were an alternative to costly public works. But an idle army is a dangerous instrument that can cut back against the man who wields it. Logically and economically, if it is to be kept employed the most profitable field of activity is against other intensive areas whose conquest pays immediate returns.

Moreover similar commercial interests are an inevitable source of rivalry. And so the great centers of intensive agriculture, commerce and civilization have throughout their history kept themselves whipped down by costly wars. The casual, sporting clashes of sparsely settled primitive man became the grim, methodical, business of mass warfare. Yet the function of combat remained what it had been, a method of maintaining biological and cultural equilibrium. War is a symptom, not a cause; yet so spectacular and terrific a thing is it, that until recent times even intelligent historians have been absorbed by it. This notwithstanding the fact that a history which is a mere chronicle of wars is like a medical treatise on typhoid which ignores the problem of bacteria in food and drink.

The intensive civilizations were not the only factor in a frequently disturbed equilibrium. Whatever the nomadic cultures may have lacked in wealth and industrial resources was frequently more than made up by their mobility and aggressiveness which we have already mentioned. As their populations increased they were exceedingly sensitive to shortage, a condition frequently made critical by the fluctuations in rainfall which are characteristic of the grassland climate everywhere.

Tempted by the great wealth of valley and maritime cities, accustomed to raiding and land piracy, nothing could be more natural for these pastoral peoples than to enlarge their operations when chance or necessity favored. From the center of continental Europe and Asia conquering hordes have swarmed out into a series of concentric fringes many times in the course of human history. However absorbing the chronicle of the Mongols under Ghengis Khan, or of the Huns under Attila, one cannot rightly interpret either without looking upon it as a special case of a profound environmental relationship. By the same token, the capitulation of the financial East to Kansas and Idaho at the Cleveland convention of the Republican party in 1936 was something more than an accident of politics. Nor is the end in sight.

We shall in later chapters have something to say about the rôle of warfare. For the present it is enough to point out that on the barbarian and civilized levels—until modern times—it has often served as a means of cultural diffusion and thus in some measure compensated for the suffering it has entailed. This of course has been in addition to the strictly biological effect which it has on the savage or primitive level of culture.

Migration and the competition of cultures are not the only consequences of man's manipulation of his environment. As populations have grown, they have frequently outstripped the resources of hygienic ritual, always a definite feature of a balanced and organized culture. The consequence is disease, often frightfully devastating. This may be in the nature of direct contagion, as smallpox; it may be transmitted by insects, as plague or yellow fever; or it may work in sinister fashion by contamination of food and water, as does cholera. In any case the difficulty of its control is increased with the massing of the human material upon which it preys.

Famine, again is a constant threat to overpopulation. Even though actual starvation may not occur, food shortage has extremely serious consequences. Many of these may not be so much due to lack of quantity, as to the lack of essential constituents. There is reason to believe that the effects of such malnutrition may show, not only in lessened vigor and increased susceptibility to disease, but upon the birth rate as well.

In the absence of major wars, then, particularly in isolated cultures, disease and food shortage may throw society into a condition of stasis—suspended animation. High death rates and lessened survival of offspring may serve to maintain equilibrium of a sort, without hope and without much visible achievement, material or spiritual. Some measure of conscious restriction of the birth rate which suggests itself as an alternative, is known to occur among simple cultures. But even then it is rare as compared with the expedients of abortion and infanticide.[1] The sacrifice of infants to Moloch was less the deliberate invention of bloodthirsty parents than an attempt to invest with religious sanction a horrible necessity.

We have traced in the most general fashion the adventure of the human animal from his lost beginnings until well toward modern times. It must not, of course, be supposed that the chronicle of any single culture or race will illustrate all of the steps we have outlined. For many the course of change has been telescoped; Japan has hurtled across centuries of cultural accomplishment in a single lifetime. Others have been caught in some remote backwash to remain unchanged for uncalculated time. Others have slipped backward, losing even the memories of their ancestral greatness, unable, like the modern Maya, even to read the inscriptions of earlier days.

But we have seen that mankind has moved through long centuries from a creature whose culture utilized his natural environment without greatly disturbing it. We

[1] Contraception among primitives may be far more efficient and war much less so in regulating primitive population, than I had supposed in writing the foregoing. For these suggestions I am indebted to Drs. Alexander Lesser and Gene Weltfish of Columbia University. For further discussion of the cultural role of domestic animals, see the author's *Lands Beyond the Forest* (New York, 1937).

have seen him grow mighty through selecting and favoring those things which he needs and desires, destroying those which he does not. Yet we have seen that he has not escaped the biological penalty of having to compete with his own kind.

Often wise and experienced, he has nevertheless throughout this magnificent period worked with the forces of nature by rule of thumb, by trial and error. His knowledge and control have been empirical, not scientific. He has not understood the laws whose operation he has invoked. Today, in large measure, he does; he is now responsible for his own acts in a sense that never prevailed before the day of science. But before we look further into that matter it will be necessary to consider other features of the culture pattern, particularly the communities, of man.

ADIOS! I MUST LEAVE YOU HERE. IF I STAYED I MIGHT ONLY MAKE MATTERS WORSE. AND BESIDES, I AM VERY, VERY EMPTY.

Chapter 15

SOCIETY, THE COMMUNITIES OF MAN

BOTANISTS are supposed to be inoffensive fellows, amiable and not very obstreperous. Suppose we take one, blindfold him, and whisk him away to some corner of the earth he never saw before. If he is even reasonably good he can look about him when the bandage is removed and shortly tell us many things. He can tell us the approximate rainfall and when it comes, something of the play of the seasons and the range of temperature, the height above sea level. He might even have some suggestions as to how mankind could best prosper and be happy in the region, although of course we would not have to listen seriously to this, for no one ever has.

If he were as knowledgeable as were Professor Tansley of Oxford or modest Professor Chrysler of Rutgers, he could name the continent on which he is placed, perhaps even suggest the nearest consulate to call upon. All of this he would tell by looking at the vegetation about him, recalling the names of plants and noticing the communities in which they grew side by side.

Suppose, however, the magic carpet drops one of us to a forced landing, not in the countryside, but among people in a village or town. We would look at their faces or their naked bodies, of course; but really to get our bearings we would look at the things they have made: their houses, their clothing, weapons, tools, and boats. People compose the community, as trees the forest; but the genius of the human community is its culture. That is the thing which sets it apart from other communities. Yet running through all of these groups, in spite of their infinite diversity, are common threads, so that we shall have to study, as we always do when working with living things, resemblances and differences, too.

First, however, let us say a word or two more about two very important matters, race and language. Just how deep and ingrained the differences between the races of mankind may be, no one can yet say. We are pretty certain that the Negro can stand physical effort in the heat of the tropics better than the white man. So far as measurable intelligence goes, every race produces superior individuals, and there is no advantage when we compare the average Mongolian with the average Caucasian.

How far are the philosophic calm of the Orient and the bustling enterprise of the Occident inborn, and how far the products of culture? The bodily odor of the European is said to be even less pleasant to the Japanese than is that of the Negro to the white; but no such odor is pleasing to the fastidious of any race, and the Japanese are notably clean. When the European exchanged measles with the Indian for syphilis (if he did) it was a bad trade both ways. Each had a measure of immunity for the disease to which he had been accustomed, and none toward the new. But this immunity had probably been acquired in the fires of time and experience, rather than being something innate. Certainly the modern Indian has less to fear from measles than did his ancestors whom it decimated.

As we have said in an earlier chapter, all of the races of mankind can interbreed, and have done so. None has an unchallenged claim to ancestral purity. The occasional dark-eyed, dark-haired Swede who surprises you is the expression of ancient Tartar invasion, and no one knows the meaning of occasional children of undoubted European stock who have distinctly Mongoloid features, often accompanied, unhappily, by subnormal mental traits.

The white-Indian hybrids of North America are no whit inferior to the general run of either race, where they have grown up in a decent environment. The only earmark they possess, if any, is a certain shyness. But good manners among Indians have always included reticence; and ten generations of living under a cloud of spiritual discouragement and defeat may well have left their mark, too. The Jews are neither a genetic nor a cultural unit, except in so far as the possession of a common code of hygiene, family life, and religious standards has kept them from deterioration. This has enabled them to maintain vitality and capacity in spite of economic hardship; but any talk of their being a united, coherent menace to western civilization is sheer hysteria based upon ignorance of their actual cultural diversity. Even their aptitude for business need involve no mysterious sixth sense. Wherever a strong tradition of family life persists for even a few generations, the young learn to avoid the errors of the old, and wisdom accumulates. So we must leave the matter of racial inheritance with the burden of proof upon those who would make it a prime factor in explaining human culture patterns.

So much, then, for the question of race; now for a few words regarding the matter of language. This is the thread by which experience is embroidered into the fabric of culture. We have shown how action and experience are registered in the central nervous system. Language is the means by which such records are resurrected, compared, and passed on to others.

The struggle of an individual for education, and the slow, painful battle of a culture towards civilization are to a great extent a problem of finding words and refining their significance. A young teacher frequently thinks that poorly trained students of strong native intelligence can be made to grasp an essential idea or relationship. As the years go by he comes to realize that a keen mind without the sharpened tool which the right word provides is as helpless as a master carpenter with his bare hands. Any education which blurs and slurs, which does not reverence the beautiful precision of words, is a betrayal of all the human mind has struggled to achieve.

French culture took over a language whose meanings had been forged in the heat of a long civilization. For centuries the process has been continued, until today the French language stands almost without a peer as a superbly tempered, beautiful instrument of precision. Whether a Frenchman has much to say or not, he has the means to say it perfectly and with the greatest clarity. German, on the other hand, and much of our own tongue, is not long away from the farm and hearth; and both still possess the very dangerous capacity to stir the heart, often before they enlighten the mind. Thus language is not merely the indispensable tool of human culture, but a remarkable index to its maturity.

To understand even a simple human community is not easy. We can thank the modern anthropologist for teaching us one lesson: no community can be rightly understood from without. Even so great a scientist, and humane a person, as the peerless Huxley failed to appreciate this fact. While naturalist on H.M.S. *Rattlesnake,* exploring remote southern seas in the middle eighteen hundreds, he had an opportunity which comes to but few men of scientific training, that of observing primitive people who had never seen a white man. Aside from making a sketch or two of their sea craft, he seems to have judged them largely by his ingrained British standards, and thought of them as quite impossible.

Man, like all other living creatures, forms communities of his own and these are a part of larger communities. Man, to begin with, must unite with his own kind to reproduce and to care for his young during a relatively long period of juvenile helplessness. This is no matter of chance, but is based upon deeply seated urges which are as potent in their way as his need for food, drink, and sleep.

The family, then is the basis of the human community. Its importance and reality are not to be disguised by the many forms which it may take, nor by the fact that sexual instinct, deep personal attachment, and reproductive activity are three different phenomena. Social pressure, romantic teaching, and religious doctrine all strive to unify them as much as possible, but the fact remains that they can be separated, and often are. Fidelity is a precious virtue because it is really a rare and difficult achievement, and not, technically speaking, because it is the rule among mankind. The family of today is constructed of physical and psychic units developed under conditions very different from those of the modern world.

Biologically man has survived and increased because the human male possesses an overwhelming and none too discriminating desire to impregnate the female. By spiritual discipline he can achieve an attitude where union with any but the woman of his choice is repugnant. Society, however, where it concerns itself at all with this matter, is realistic, and takes no chances. It employs many practical safeguards, economic, legislative, and social.

On the other hand, mankind has succeeded biologically because desire in women, whether transient, periodic, or constant, is replaced after impregnation by an equally overmastering urge, the maternal complex. Science is none too clear even yet as to the interplay of these two great forces in the human female, and biological experience would lead us to expect no simple formula. One thing is clear, however, and that is that where the maternal impulses of woman run counter to the sexual urge in man there occurs, and probably always has occurred, a powerful divisive force in the family, fundamental unit of society though it be.

In the family, too, there has always been, of necessity, a division of labor. Man's part in reproduction leaves him greater mobility and more muscular freedom. In common with males generally, he is better adapted for combat with other animals and with his own kind. Biologically he can assume greater risks since parity of numbers between males and females is not essential to reproduction. Although infanticide may select girl babies as its victims, in warfare the lives of males are freely hazarded to protect adult females and their living young.

With the advent of modern technological culture the division of labor within the family has become more, rather than less complicated. The lengthening chain between producer and consumer is scarcely offset by the greater abundance of goods, even of food. In plain language, food is no good unless one has money to purchase it. Children represent a prolonged burden to both parents and it is doubtful whether social measures, such as schools, are as well fitted for their task as were simple tribal custom and ritual. Harsh though the latter may seem they at least prepared the young for the kind of life that had to be faced. The family of today encounters disintegrating forces far beyond the basic emotional discrepancy of biological instinct we have mentioned.

All of these problems admit of scientific approach and impersonal study, it is true. But intense personal issues must be met in solving them. It should at any rate be clear that no simple answer, like that which can be obtained in the transmission of electric power, or the conversion of crude oil into a wanted drug, need be expected. If family stability is as desirable as reason and experience indicate, it will be best guaranteed by conditioning the young to maintain it. But with an entire world in flux, it is practically impossible to foresee, in one decade, the things which children must be conditioned to meet in the next.

Beyond the family, complex as we have seen it to be, there exists a vast range of more elaborate communities. These can be and have been arranged into a very pretty series—clan, tribe, race, nation; with their various administrative expressions—hamlet, village, town, city, and state. But such classifications are too formalized. They are expressions rather than the basic stuff of the cultural web.

The unifying principle in the study of human cultures is the analysis of the patterns they form. These patterns are built of culture traits and complexes of most diverse origin, by no means always native to the region or group which exhibits them. They have developed, been transferred and modified to meet the survival needs of the group as these present themselves in a given environment. Food, shelter, clothing, entertainment, sex, care of young, war, disease and death, order within the group—these form the inevitable framework behind any pattern when we examine it long and sympathetically.

In our own culture pattern today we know vastly more about the influences which affect our daily needs than does any savage. Certainly we know they are many and complicated. It is our idea to understand and control them if we can, or adjust ourselves to them if we cannot. With all of our science we succeed but imperfectly.

In this we are not so different from the primitive as we suppose. In more direct and immediate contact with the physical world than are we, and in general better insulated from alien cultures, the primitive is decidedly regional. There is greater uniformity of dialect from Pennsylvania to Oregon than between adjoining counties in England, the English dialects being a heritage of the days of much more restricted movement and contact than the present.

The cultural responses which a given region will evoke are largely unpredictable. The human animal is too resourceful, and he enters a new area with too variable a heritage from his previous experience. The Kickapoos who have returned to Oklahoma from Mexico insist on roofing their huts with rushes, while the nearby Shawnee use the boughs of trees for their summer shelters; and the one-time derelicts along the Canadian River chose neither, preferring sheet metal if they can get it from city dumps.

Regionalism, while wider in its scope in the modern than in the primitive world, remains a distinctive feature of sophisticated culture. New York is, despite its cosmopolitan population, an intensely provincial place. The Pacific Coast is as exercised over possible war with Japan as the East Coast was inflamed over the World War and the Middle West initially apathetic about both. Radio, that supposedly universal means of communication, finds itself obliged constantly to take into account marked local variations of taste and interest. The boundary between cotton and wheat in Oklahoma is marked by differences in politics, religion, architecture and economics. Great mineral, forest, and climatic centers in North America have developed distinctive patterns—often confused and overlapping, of course, yet readily perceptible.

In addition to regionalism, which is an ecological factor certainly, culture patterns exhibit another feature closely connected with it: the response of man to the problem of unknown forces. The modern world, as we have seen, approaches this problem through scientific technology and is inclined perhaps to ignore the limits beyond which that method will not operate, as well as to neglect its ultimate consequences. This has resulted in exploitation and a sort of cultural anarchy which will be terminated by its own effects if we do not see fit to act before it is too late.

The oriental world approaches the same problem of spiritual adjustment through passive resignation to suffering, the acceptance of hazardous living standards and of rigid social barriers. The western world of the Middle Ages approached it through the church, which at least embodied law and hope.

As to the primitive cultures, they appear to have responded by creating patterns of ritualism, taboo, and magic. Absurd when viewed apart from their general setting, these patterns of response exhibit a rigorously logical structure within themselves. Holy Church has no monopoly on consistency. In primitive cultures every phase of existence is permeated by patterns of rigid convention, bespeaking prolonged effort to adjust the social group to the world in which it lives. Symbolism plays a large part in primitive life, but so it does in civilized life. What is the rosy-cheeked lad in the soup advertisement, or the languishing bathing beauty in the motor car folder, but a symbol, as potent to us as the circle or swastika to a Pueblo Indian? It is scarcely necessary to remark that the cultural adjustment to unknown forces is as unpredictable and as varied as the pattern of practical adjustments we have already discussed. In short, human cultures must be studied and understood by watching them grow and function rather than by attempting to calculate their proper orbits. Science must content itself with discriminating and intelligent retrospection.

In this connection it is interesting to recall that when Darwin, in all modesty and caution, propounded his theory of organic evolution, it was attacked by the logicians of Europe because it was not a structure of pure, cool, impersonal reason. It was rated as a vast example of special pleading, with evidence marshaled from special cases to fit the hypothesis. But the world has since come to learn that no broad and synthetic attack upon any problem involving life can proceed with the mechanical perfection and smooth deduction which are the delight of mathematics and her handmaiden, theoretical physics. Mother Nature herself provides the often elaborate assumptions from which deductive theory must start, and the task of the student of life is largely to study and record the given conditions. As soon as he selects, or like the mathematician, invents them, he faces trouble. The failure of even so profound and accomplished a naturalist as Aristotle to realize this probably delayed the advent of modern biology for eighteen hundred years. It is the fate of the biologist to spend weary years in unifying, interpreting—rationalizing if you will—a multitude of complex actualities before he can achieve the right to calculate expectancies or put forth the most modest prophecies.

In human relationships the idea of culture patterns is an example of a working truth arrived at by such a method. We have seen that the culture pattern stands between the individual and the physical environment. In other words, the culture pattern becomes the immediate environment of the human being and the medium through which he reacts upon the physical world about him.

The importance of this fact cannot be overstressed. The hard-headed realist may call this nonsense. Does he not feel the wind and rain upon his cheek, and stub his own toe against the rock? Is not his contact with the environment an immediate, direct, and personal thing? It is not and cannot be, even if he retires from his fellows and lives in solitude. He cannot think of climate, soil, or his fellow creatures without employing the words and mental pictures with which his culture pattern has furnished him, even when he withdraws from it. The wind, the rain, the pebble, and the tree can never be present to him as mere physical experiences. All are highly conditioned experiences, and the conditioning comes from the prevalent culture pattern.

Poet, scientist, and business man, to take three extreme types, are all accustomed to believe that they bring to a new physical situation an unusual independence and freedom of attitude. The poet responds to a landscape or to a personal relationship with phrases whose meaning and emotional content are the most intimate and sensitive expression of the culture in which he finds himself. The business man is incapable of thinking in terms apart from the economic order in which he finds himself—as a business man, that is to say, for every human has his various facets. The sturdy scientist, hewing to the line, regardless of consequences, is in some respects most highly conditioned of the three. He cannot even think without employing symbols, terms, and relationships which have been standardized, registered and officially approved by competent authority. Compared to the calculated and deliberate intent—the consciously cultural intent—with which science has been fostered the growth of poetry and business have been spontaneous wildings in human society.

It may seem unnecessary to carry the point further, but it will not be uninteresting to do so. How does the physical environment operate upon the average individual on the street or farm? Always through the culture pattern in which he has been reared and which interposes itself. Changes in temperature become significant to him in terms of weather reports and oral tradition. Their influence upon him is cushioned or intensified in terms of the clothing, the diet, the markets, the architecture about him—all expressions of his culture pattern. Experience apart from the surrounding culture pattern does not exist, perhaps not even for the insane.

We have yet to consider another aspect of the culture pattern, that is, its importance as the means by which man acts upon the physical environment. This is a matter of the utmost cogency when we recall what has been said about man as a dominant organism in the world community today.

Everywhere we turn, whether to primitive or highly urbanized human communities, we find that human action upon the environment no less than its reverse, is conditioned by the prevalent culture pattern. The environment, in fact, merely serves as a means to keep the culture pattern going and is exploited without limit for that purpose. Invention and the use of substitutes arise when environment no longer yields what is needed for the culture, but the old patterns persist with the new material, as buggy wheels and tops lingered in the automobile industry until they interfered with the new speeds and strains which were made possible. It is a characteristic of the culture pattern once launched to demand its own expression at any cost whatever to the environment.

This, of course, cannot go on forever without penalty. Yet so great is the power of the pattern under which we live that it produces in society inertia and conservatism, and keeps us from looking ahead. The individual who sees the coming of difficulties finds himself pitted against the remorseless workings of the culture pattern. The voices of those who urged caution in removing the forests of this country, while forests still remained in abundance, were drowned in the roar of ceaseless exploitation about them. Today as we face the certainty that our great oil reserves will some day be depleted, only a fool in his folly would hope to stem the rush to remove oil from the ground as fast as modern applied science can accomplish it. All of the great continents bear witness with abandoned cities, mines, and fields, drifting dunes and denuded hills, to the fact that human activity can persist in a suicidal course of doing the things that it has always done, in the way that it is accustomed to do them until the means of survival themselves are exhausted. Wherever intelligence becomes polarized into routine, the human animal is in danger of doing what the chestnut bark blight fungus has done to itself by destroying all of the native chestnut trees in the eastern United States.

There are, however, other possibilities. Primitive societies exhibit many instances of a balanced relation to the environment. These are usually marked by practices which restrict population growth, by frugal standards of living, and use of the means of support, and very generally by huge reserves upon which to draw. There is perhaps no better illustration of such a balance than is afforded by the buffalo-hunting culture of the plains Indians. Severe and hazardous, the mode of life served in itself to restrict population. A technique of very thorough use of the buffalo carcass and of preservation of the meat was combined with a wide knowledge of edible, self-producing native plants, and a modest, somewhat casual, agriculture. An ingrained sense of dependence upon nature had developed into an abhorrence of waste and wanton destruction. The absence of any sense of private property prevented anything like commercial exploitation. And the vastness of the natural range, with its shifting herds of buffalo furnished the necessary reserve, the margin of safety within which the culture pattern could adjust itself.

Similar examples of balance, safeguarded always by huge reserves, are to be found in the agricultural areas of Abyssinia, Egypt, and China, and in the fishing, fruit-eating cultures of the South Sea Islands. In the agricultural areas mentioned, the prevalence of a primitive type of hand labor serves both to keep the population busy and contented, and to prevent wasteful mining of the soil, while great rivers which rise in scantily populated regions contribute the reserves of fertility and moisture which are essential. Modest standards of living contribute, as among the plains Indians, to retard exploitation. In the South Seas, the ocean itself furnishes the reserve, which is tapped by daily fishing.

Alternative to balanced adjustment is migration. This may be due to periodic depletion of resources, with a resulting pattern of movement among the cultural group concerned. The familiar trek to summer hunting or fishing grounds, pastures or garden spots, and the return to winter villages is an example of this, and frequently represents a special type of balanced economy.

On the other hand the depletion may be more serious, as among the forest burners of Africa, Northern India, and Central America. Crude agricultural practices under conditions of rapid soil exhaustion or uncontrollable weed growth may make it impossible to remain permanently in one place. A British colonial administrator, returning to take charge of an area after an absence of some years, found it depopulated and the soil ruined by erosion. Before the British had interfered with warfare and primitive sanitation in Africa the social groups were not large or numerous enough to create a serious problem under this type of existence. But with life safeguarded and pressure added for commercial production from the soil, a problem of first magnitude has arisen. There is no longer the old opportunity for environment to recuperate before the migrating groups are forced back to it.

Throughout history and prehistory, of course, more or less permanent migration has been forced upon social groups which, by the character of their culture pattern, have despoiled the environment. Often the crisis is precipitated by more or less periodic natural events, notably drought. One does not need to revert to the great waves of nomadic movement out of the European steppes to exemplify this; the population shifts in the western states due to recurring drought and more recent rural-urban shift are sufficiently clear. Needless to say, migration, however caused, has been one of the means whereby man has achieved his present terrestrial dominance.

A further alternative to extermination, adjustment, or migration is that of invention. While obviously an effort at better adjustment, it has had the net effect of destroying, rather than establishing equilibrium with the physical environment. A people employing stone tools are under very definite limitations in securing materials, whether by digging, chopping, or piercing. The replacement of stone by bronze, later by steel, remove these limitations, it is true, but in doing so create many new problems. The invention of agriculture and the whole host of inventions which are comprised under the invention of the scientific method will presently be discussed in this connection.

Man no longer suffers under handicaps of mobility or of technique in converting the environment to the purposes of his culture pattern. He uses it freely to destroy what is left of more primitive, better balanced patterns. To date modern man has been singularly successful in finding his way around problems created by depletion of the environment, although some of this success is certainly illusory. Even the boom time production of 1929 is estimated to have fallen about 80 per cent short of producing the goods necessary to insure reasonable comfort and satisfaction to the American people, as judged by present standards.

Everywhere today is felt increasing, critical pressure of the modern culture pattern against the environmental reserves on the one hand, and against the individual on the other. The difficulties of adjustment are intensified by the very complexity of the pattern, and by *the inability of one who is within it to view it in any terms save those which it has afforded him*. Will it, in the manner of an organism to which it has been so frequently likened, proceed spontaneously to new heights or new depths? Or can reason, so often a mere instrument in the elaboration of its details, become a factor in its shaping and control? No one knows. There is food for thought in the fact that the Russian Revolution, whose seemingly insensate, fatalistic, irresistible fruition has been so marvelously described by Trotsky, is very different from what it might have been but for the years of careful, devoted intellectual work which preceded it. No matter how one feels about its results, there is a powerful ray of hope for mankind in the manner of its shaping. Although man is caught in the web of his culture pattern, yet he has the right to try to modify it—if he can.

Chapter 16

DISEASE, THE FAILURE OF ADJUSTMENT

"NEVER trouble trouble till trouble troubles you." Thinking of disease, as most laymen do, in terms of a purely personal hazard which involves a large element of luck, few of us concern ourselves with its wider significance in nature.

And in response to this attitude, the training, code, and economics of medicine have been largely built upon the assumption that disease is chiefly a personal matter. Of course the traditional physician has been very much preoccupied with the immediate work in hand. Yet it is rather depressing, in spite of the brilliant advances of the past century, to see how slowly the social importance of disease has won a hearing.

True, diseases whose social cost is obvious and staggering, or whose prevention, like that of smallpox, is relatively a simple matter, have received attention. Brusque generals whose fighting power is threatened by contagion, are not likely to mince matters when it comes to securing results in social, rather than personal terms. Life insurance companies who are in a position to measure—and pay for—the ravages of disease, are conscious of its importance to the group as well as the individual. The growers of domestic plants and animals have their means of livelihood threatened by diseases, and have learned that this threat cannot be met by isolated efforts.

Today most communities have health establishments of some sort. Knowing this, we too frequently relax as though the task were done. Oftener than we know, such arrangements represent little more than a feeble gesture. The powerful efforts of great foundations, and the heroic labors of devoted individuals still encounter, not only apathy and inertia, but downright frustration through political and financial selfishness. And our attitude towards venereal disease has been, until very recently, nothing less than barbaric.

Disease is a disturbance of normal behavior or function, and not primarily a matter of structural change. To use an analogy, we might rearrange the brick and other material in a house, even remove a good deal of it. The result might be grotesque and bizarre, but until it reached the point where the house could no longer function safely and effectively as a dwelling, we would not condemn it. But the house whose foundations and walls will no longer safely carry the roof, or whose roof no longer turns the weather, would, by our analogy, no longer be in a condition of health.

Usually the term "disease" applies to living organisms only. But it applies to all of them—plants as well as animals, microörganisms as well as large and conspicuous forms. All of them are subject to it, and the principles involved are remarkably general, but this fact is not easy to uncover behind the wilderness of detailed symptoms and individual causes. The living world, be it remembered, consists of over a million known species, many of them in their own right sources of disease to their fellow creatures. Moreover, diseases have been cataloged and classified from time immemorial on the basis of their symptoms—that is, their visible manifestations—which is a very different matter from the cause, or the real disturbance involved.

The work of Pasteur and his successors threw into sharp focus for a time the importance of bacteria and other microscopic life as a causal factor in disease. Within a few decades he made history as few mortals have done. It was a momentous occasion when he demonstrated that organisms present in diseased tissue could be isolated, grown, used to produce the disease, and then be recovered from the body of the new victim and grown and used again. From that time on there could be no quibbling against the old proverbs about cleanliness and prevention. The provisions of Leviticus, that remarkable code of Old Testament law, could no longer be regarded as the trivial whims of a jealous and captious tribal god, but as an exceptionally good use of practical experience in dealing with unknown forces.

Equally momentous with the discovery of bacteria as a cause of disease was the subsequent discovery that immunity could be acquired. The means utilized might vary—dead bacteria as in vaccination, weakened strains of living bacteria, or serum from the blood of an immune animal—but the principle involved is that of producing a chemical change in the host which would thenceforth render him immune, for a shorter or longer period. In 1903 the armies of alert Japan were thus virtually protected from diseases which had laid low the American troops just five years before.

The influence of nutrition upon health and disease had attracted the attention of the earliest medical scholars. But nutrition itself was a mysterious subject until the great development of organic chemistry during the nineteenth century, and particularly toward its close. With a better understanding of the composition of organisms and the food which they require the matter of a properly balanced diet as an essential factor in health became clear. And yet it must be remembered that just as we have in the Jewish law a very excellent hygienic and sanitary code, so in the dietary practice of most stabilized human cultures we have an excellent and wholesome combination of essentials to normal nutrition. Our own western civilization which had scarcely recovered from the environmental ills of the Middle Ages before the Industrial Revolution was upon it, has not done so well in this respect, at least not until very recently. Thanks to the cleverest guild of dentists in the world, Americans have largely been able to conceal the ravages wrought upon their teeth by too refined a diet. The Europeans, as every traveler knows, have not been so lucky.

Within the past two decades it has been abundantly proved that a mere balance of proteins, carbohydrates, and fats is not enough. Not every protein has all of the constituents necessary to normal growth and development; for example the proteins of maize are deficient in this respect. It is a saying among farmers that cattle will starve to death on an abundance of maize, if they have no other source of food; which is the farmer's way of saying that they will look ill-nourished. And this is not from lack of protein, but of the proper kind of protein. Add wheat or alfalfa and the need is met.

More striking perhaps, certainly more readily commercialized, have been the series of discoveries explaining the nature of rickets, scurvy, and other discoveries of growth or of nerves, often occurring where food is plentiful but lacking in variety. One of the classical true stories of the sea describes an attack of scurvy among a ship's crew and its quick relief when the men secured large onions from a passing Portuguese vessel and munched them like apples. This was due to the presence in the onions, as in other fresh vegetables and fruits, of one of the several vitamins which are essential to a proper animal diet. Of plant origin, these vitamins may be found in certain animal material—as cod liver, for example—in concentrated form; some are soluble in water, others in fats; not all are equally stable, none more than moderately so. Great progress is being made in the isolation and study of the vitamins and grave disturbances are now readily prevented by the easy method of supplying them.

Two other developments have served to broaden the basis of our attack upon disease, each of them opening up new vistas almost comparable with the kingdom of the microscope which Pasteur made accessible. Viruses and glandular secretions, or hormones, are now known to be efficient agencies in the matter of health and disease.

It happens that a number of diseases, notably measles in man and mosaic in the potato, are clearly transmissible, yet have never shown the presence of a visible causal organism. Whatever causes the disease increases within the host, just as do bacteria, but it cannot be seen under the microscope, as they can. It may be extracted in fairly large amounts from the diseased tissue, along with other body fluids, and is thought to be proteinaceous in character. It is very similar to living material in its sensitiveness to poisons, temperatures, and drying. And for lack of a better name, it is called a virus. Those viruses which cause plant disease—and there are a number of serious kinds—are often carried about in the mouth parts of insects.

Equally difficult as to their composition and mode of operation are the hormones or glandular secretions. Not only does it appear that the health of an organism is affected by the substances which enter it from the outside. The balance between its own products may be quite as serious a matter. Much of our ability to judge people by looking at them is based on the fact that these hormones influence both structural and psychic traits. Not so many years ago the author of a book on *Glands and Personality* would have counted for his audience upon the scientifically illiterate; today his general thesis, if not all of its details, is scientifically respectable. Like viruses, hormones appear to affect plants as well as animals—grotesque patterns of growth can be caused by local application of a suitable hormone to the side of a tomato or maize plant. Among animals one of the most interesting hormone effects is the so-called "free-martin." This is the female of a pair of twins of opposite sex, generally sterile in mature life because of the chemical, that is, the hormone effect from the accompanying embryo male.

In the preceding chapter we have mentioned the often deleterious effects of foreign proteins, whatever their source, upon certain sensitive, or sensitized hosts. This effect, best known in food rashes and hay fever, is known as allergy. The list of allergic diseases is constantly growing, and there can be little doubt that allergic phenomena will furnish an important clue to any ultimate theory of disease. To an amazing degree, life is the expression of relationships within and between protein material, for of such composition, largely, is living stuff itself. The purely allergic diseases differ, of course, from the virus diseases. In the former the trouble ceases as soon as the source of foreign protein is withdrawn. In the latter, the foreign protein, or virus, once introduced, increases in amount and persists. Whether this is through its own direct chemical activity, or because it stimulates the living stuff of its host into a new activity, is a very important question indeed. Nor should it be forgotten that no parasite can be long successful in the presence of a host whose proteins are incompatible to it.

Thus we arrive, not by a study of so-called normal behavior, but by an examination of the abnormal, or pathological, at a new and important factor in the interrelationships of living things. We have seen the efforts of the entire organism to compete for light, water, minerals, food, and space. In that struggle its weapons are afforded by the structure of its bodily organs, and by the behavior pattern of the whole individual or of the social group to which it belongs. In that sense we are dealing with adjustments on the grand scale. But in this matter of chemical compatibilities and influences we see adjustments of a far different order of magnitude. Here the scale involved is that of the chemical molecule. At present much more difficult, technically, to approach than the more evident adjustments we have mentioned, one may venture the guess that their basis will be found to be electrical and structural, that is, largely physical. And while we need look for no ultimate explanations in either the field of the whole organism or its constituent molecules, a working interpretation of the latter that is comparatively simple will probably be forthcoming. The really elusive mystery will remain in that tangible, obvious thing with which we can at present work so much more easily—the living, organized plant and animal.

There will lie, after the mechanics of disease is understood, the real problem of the meaning of health and disease. Human disease we regard, from the personal suffering it causes us, as an unmixed evil. Society commits itself, rightly enough, to the elimination, or at any rate the mitigation of this evil, both in man and his domestic plants and animals. As to the latter, whose numbers can be more or less satisfactorily controlled, there can be little question. In the case of forms of life inimical to man, disease is our welcome ally, of course. We cheer for the white rust of crucifers when it destroys the troublesome wild mustard and shepherd's purse; we are considerably less enthusiastic when it attacks our garden vegetables belonging to the same family. And when it comes to the great majority of organisms which grow wild in nature, neither directly useful nor harmful to us, we see in disease an important factor in regulating their number.

In the absence of disease, of course, the food supply would do this. But probably it would be at the expense of periodic depletion of plant material, even granting that predatory and carnivorous animals could be freed from the toll of disease. There is, in fact, a close connection between these two means of holding life in check. In one of the national forests, where the preservation of deer seemed desirable, wolves and coyotes were eliminated by poison and shooting. Shortly the vigor of the growing herds of deer began to decline. Misshapen and abnormal specimens became numerous and of course, exerted an effect as breeding stock. The restoration of the wolves became necessary, curiously enough, for the well-being of the deer. Nor should it be forgotten that the wolves, besides starvation of their young and the ordinary vicissitudes of existence, are subject to parasitism and ill health. For a sick wolf there are no bed trays and kindly nurses. And a hungry wolf has to catch what he eats, thus, at one stroke, exemplifying the best features of both the merit and the spoils system.

In the human race, the problem of disease cuts across many aspects of life. In the days when a nation's working power was an asset instead of a nightmare for statesmen, the loss of effectiveness from ill health was a serious economic problem. Of course it still is, even in our mechanized world whose great problem is the use of idle time. Unquestionably, too, the care of the ill and weak gives expression to some of the finest attributes of humanity. But it can also evoke some of the worst, including commercialism and that most stupid and cruel of traits, sentimentality.

Until the rise of modern medicine, disease was as potent as famine and warfare in regulating population growth, and the three were recognized as dreadful alternatives. War, until it assumed its modern, impersonal horror, was perhaps the best of the lot, for it at least gave the chance to die while trying. As to a famine, while there is no longer any excuse for it, it still, as we have said elsewhere "stalks the land in the invisible cloak of malnutrition." Disease, too, has changed in the character of its incidence. By conquest of the more prevalent epidemic types the average length of human life has been greatly increased, and more lives are terminated at a later period by such baffling and complex disorders as cancer, kidney, and heart disease.

We do not know the biological effect of thus prolonging human life. There is no proved correlation between the social value of a man and his immunity, say to measles or smallpox. The roster of tuberculars certainly includes some dazzling names, and the same may be said for many another disease. The records of a great hay fever and asthma clinic indicate that the majority of their patients are of superior mental and physical vigor. But the fees in that clinic are substantial, too, possibly a selective factor. Eminent British scientific opinion holds that the efficiency of modern medicine is impairing the heredity vigor of the race; but that argument is probably as old as the shift from foot to horseback and the change from flint to steel. Selection will go on, whatever the environment. The basis merely changes with the conditions.

Much more serious, immediately, is the change in the composition of the age groups in society. Modern industry will not employ men past middle life, yet these men retain their capacity for useful work and not uncommonly have the maximum responsibility for dependents. A master machinist of the old school was frequently useful until seventy; it takes youthful vigor to keep pace with modern machinery. Quite apart from any emergency problem of unemployment, this group of elderly unemployed, steadily increasing at a time when the penalty of a disordered physical environment hovers over society—a bolt about to strike—is something we are yet to hear the last of.

Yet, while we need vastly more information than we have of the social effects of disease and its control, it seems likely that we can do little better than assume that human diseases in general are more disgenic than the effects of their prevention and cure. Common kindliness dictates that belief, and it has been suggested more than once that kindliness—not sentiment—may lie fairly close to the heart of human cultural problems.

Certainly the solution of the most disgenic of diseases—those that affect the process of reproduction itself—lies along that pathway. Toward the problem they afford society has chosen to be neither kindly nor, which is worse, realistic and honest with itself. The venereal diseases reap their heaviest toll among those who, for reasons largely economic, are obliged to satisfy the normal and beautiful sexual urge under sordid conditions. They are the victims of a situation they did not create, an unhappy threat to future generations and to society itself. Yet, in spite of heroic efforts in which the Federal government has played no small part, their problem is still dealt with furtively and ineffectively. If kindliness does not demand the alleviation of their misery, ordinary selfishness should see them hospitalized and given the best treatment for the protection of society itself.

There remain many aspects of disease, both in plants and animals, which are of the greatest ecological importance, for disease itself is primarily an environmental phenomenon. Among such aspects are the cases of parasites which must have alternate hosts, such as man and the rat, or wheat and barberry; also the rôle played by living organisms as vectors of disease; and the strategy of disease control—so often achieved by ecological measures. The Italians are today, for example, stocking their marshes with American top minnows, small fish which eat the larva of the malarial mosquito.

Again there is the vital problem of constant evolution among the organisms which produce disease and the seemingly endless aggression of new diseases under changing conditions. Likewise there is the growing seriousness to livestock, possibly even to human beings, of mineral deficiency diseases in great areas whose soil has been so rapidly eroded or otherwise impoverished of such minerals as phosphorus.

But to prolong their discussion would carry us quickly into the realm of the specialist and would merely at the end confirm this most essential truth: disease is an environmental force of the utmost importance, whose implications, no less than whose control, demand the attention of an intelligent society.

Chapter 17

TECHNOLOGY, SLAVE OR MASTER?

EXISTENCE is a gamble for every living thing. Man is no exception to this rule. Those who live by gambling know that there is only one hope of reducing hazard; that is through knowledge of the mechanism upon whose performance the hazard hinges. Even this is no perfect guarantee.

The spread and domination of the human species suggests that it has been slowly reducing the odds against itself. Yet everywhere struggle, hunger, and misery are present as proof that the old hazards disappear but little faster than the new social mechanism arises to replace them with others more baffling, less tangible.

Steadily man has increased the power and the effectiveness of his bare body as an instrument in the struggle against environment. Some of the devices that he has developed have been described. Of late modern science has been producing them faster than they can be utilized, and has changed the whole framework of man's living while he himself, in mind and body, remains substantially unchanged.

The beginnings of human culture have been traced back five hundred thousand years, but fifty thousand years is a conservative estimate of the length of time that modern man, as a distinct species, has been in existence. Eight thousand years is a generous estimate for the age of agricultural and urban civilization, such as it was, and is. As an effective instrument in the western world it is scarcely half so old. And, with the exception of very simple mathematics and physics, its methodical and conscious approach to environmental control is not more than four hundred years old.

Moreover, the revolutionary social and economic effects of this methodical, that is scientific, approach have scarcely been in operation for one century. During that century these effects have moved ahead increasingly, allowing no temporary respite in which to achieve more harmonious pattern integration of the ever-growing host of new culture traits. Yet it is one of the fundamental laws of ecology that an organism must make its peace with nature, on terms of even exchange, if it is to secure peace of existence in return. Meanwhile, physically and mentally, as we have said, the creature who struggles to adjust himself in this maze of new and unforeseen forces of his own creation, is largely the being who lived before the dawn of history. And thus it is, that before we consider the resultant environmental problems, we consider the means by which man has brought them about.

Throughout most of its existence the human race has been dependent upon blind experience for its lessons. This is a slow, often costly, but eventually sure, procedure. Let us imagine someone sensitive to poison ivy who has never been near the plant and knows nothing of it. He walks through a certain thicket and presently develops a severe dose of blisters. He may attribute his misfortune to any of a dozen causes. Slowly, by repeated experience he may come to associate his trouble with a certain place, or type of place, and avoid it. Much more slowly indeed is he likely to discover that his ailment is due to a particular plant growing in the kind of place he has learned to avoid. For practical purposes he has learned what he needs to know and learned it well, but the experience has been costly. Yet such dearly purchased information has been, until recently, the source of all arts and crafts—construction, warfare, agriculture, medicine, cookery, and music for examples. In many cases this empirical knowledge has made possible achievements well-nigh approaching perfection. By 1725 the violin had achieved through a continuous process of refined trial and error a degree of perfection to which no conceivable amount of modern acoustic analysis can contribute more than an explanation. Water transportation from the days of crude rafts and dugouts progressed in similar fashion until the old New England salts with their formula for an efficient ship "tail like a mackerel, head like a cod" anticipated the whole essence of modern streamline dynamics.

Perhaps no phase of environmental relation shows this principle of empirical adjustment more strikingly than does the matter of food and diet. To begin with there is no known food source whose discovery can be credited to modern man. Evidently our primitive ancestors wherever they lived explored pretty thoroughly the possibilities of the local fauna and flora. Lest this seem incredible, it is well to remember the lengths to which starvation will drive the most civilized. Moreover, in evaluating the food habits of other cultures than our own it is essential that we know these cultures and the environment with which they have been integrated. The eating of small reptiles and insects differs in no great respect from our own use of marine crustacea, white bait and live oysters, and is less hazardous to health than the consumption of doubtful milk from carelessly washed glasses in an unscreened public eating place.

Even more interesting than this early exploration of possible food sources is the development of food habits and dietary in cases where some choice has been possible. The menace of an exclusive meat diet to the modern civilized man is well known. Yet such men if they live among the Eskimo thrive as well as their hosts who have for long periods nothing but meat to eat. The answer lies in the fact that the Eskimo make much more thorough use of the carcass than we are accustomed to do. The fats, rich in hydrocarbon, replace the carbohydrates of our starchy and sugary vegetables. So far as possible entrails, glands and internal organs, marrow from the bones, and blood are consumed, thus supplying to a large extent necessary constituents which are absent in the muscle tissue—our usual article of diet. When a walrus is killed, the stomach contents consisting of freshly killed shellfish, seasoned by the sour gastric juices, are the equivalent of our delectable salads. For the clams are themselves gorged with microscopic marine plants.

There is no better illustration of the sound basis of these purely empirical food habits of the Eskimos than to consider the change in packing plant practice which has developed within the past two decades. Liver which was formerly easy to buy is now an expensive luxury in butcher shops because it is known to prevent anemia. Glands which were once converted into tankage are now carefully separated and processed, to be dispensed in costly small amounts by druggists for the remedy of unbalanced conditions, brought about by our very artificial way of living.

The effectiveness of the old rule of thumb may be still further shown in relation to the preparation of food. Natives who use the cassava root free its nutritious starch from the deadly poison which occurs with it by leaching, just as the modern chemist takes advantage of relative solubilities in his analysis. Cookery itself represents a great technological advance, fully justified by the work of Pasteur and his successors. Tea and wines afford drinks safer than untested water or raw milk in lands where cholera and typhoid are a constant menace. The fermentation of vegetable and dairy products is not only a safeguard against injurious bacteria and against decay organisms, but attained in the days before science a degree of perfection which has fully occupied the scientist to explain and duplicate. Honey, used by the Vikings in brewing mead, will not ferment because of the presence of traces of volatile formic acid. But the old recipes which have been handed down show that our ancestors boiled the honey, thus driving off the preservative added by the bees. Similarly in the manufacture of distinctive types of wine and cheese, a most exacting ritual has developed through the centuries, by the process of persistent trial and error.

The domestication and husbandry of plants and animals, that is, the technology of food supply, has already been discussed. Here, too, trial and error slowly developed improvements in practice. Digging stick, hoe, and wooden plow drawn by domestic animals are significant steps toward greater effectiveness. Mechanical cultivators—horse-drawn hoes—first designed by Jethro Tull in the early eighteenth century represent a relatively high type of achievement of mingled empirical and scientific character. This device was the result of deliberate experiment and intention to improve husbandry. Yet Tull's writings show plainly that he could not have understood just why better results followed the use of his instrument. Indeed there is some question whether we of today agree on the matter.

Irrigation as well as storage and preservation of food were carried to a high degree of perfection before the dawn of modern science. In the reclamation of ancient irrigated lands of Asia Minor it has been found that modern engineers can save themselves trouble by taking aerial photographs of abandoned ditches and restoring them substantially as they were. Drying, smoking, salting, sealing, and cold storage are as useful today as ever, with the result that we have busied ourselves to improve these venerable practices rather than to discover new.

What has been said of the arts connected with nutrition can as well be said of those connected with construction, transportation, destruction, and protection of the body and the healing of disorder. Of course due allowance has to be made for the difference in degree to which these various groups of arts could progress in the absence of modern scientific resources. The modern ocean liner could not exist without modern physical science, but the clipper-built sailing vessel was no less a work of perfection on that account. The pyramids of Gizeh and Tehuacán have stood longer than the Empire State Building is likely to. The heavy silk brocade of a mandarin will still undergo comparison with the most marvelous nylon fabric, product of refined experimental science as the latter is.

On the other hand the knowledge and skill behind the clipper, the pyramid and the mandarin's coat have cost infinitely more in human effort and waste than the added measure of knowledge and skill behind their modern successors. Wherein lies the difference? Briefly it lies in the invention of a new method, or procedure—that of discovering principles and using them to solve problems.

The simplest example of this procedure is the sort which the child, like the race, learns first to employ. Numerical measure is something which can be applied quite generally. The relations between numbers, once we have a system of symbols, can be worked out with considerable ease. The student who complains that mathematics is hard for him may well doubt that it is actually the simplest subject that he has to study, yet such is the truth. His trouble arises from the rigor and persistence with which its simple material must be developed, rather than from any lack of simplicity. For example, once we know from persistent experience that all even numbers can be divided into equal halves, we are ready to apply the principle in the case of any particular even number.

The exchange of merchandise, measuring of land, and reckoning of time by astronomical events all stimulated the early development of simple mathematics.

In slow and halting order, often with prolonged lapses, the seven seals of science, as Mayer calls them, have been broken—mathematics, astronomy, physics, chemistry, geology, biology, and psychology. At every stage new vistas of power and environmental control of a conscious, deliberate sort have opened up. And each new step has made possible further steps, not only in its own field, but frequently in others. No better illustration of this exists than is afforded by the development of optical theory in physics. By the telescope which resulted man became aware of the extent of the universe and the astronomical character of the earth on which he lives. And through the microscope he saw, literally, into himself and his fellow organisms. Scientific biology became possible.

There is ample evidence in medieval writings that those who dabbled in the evil art of experimentation sensed very clearly that enormous resources lay just beyond their grasp. The civilization of today is the inheritor of their dreams. Space and time have lost something of their inexorable character in the face of aviation, radio, and other electric devices. The stored sunshine of bygone days, oil and coal, have been harnessed and made to flower into infinitely rich manifestation. The evidence of modern technology, based upon conscious study and control, is much too near at hand to require elaboration. It is in fact, too near at hand to be comprehended under any circumstances. Yet it is worth our while to attempt to frame some sort of picture of the new relation between man and his environment in the modern world.

Perhaps the most striking effect of science has been to intensify the separation between man and his natural environment—a process already begun with the empirical development of agriculture and the consequent increase in population and the growth of cities. The modern man thinks in terms of energy and materials far removed from their source, transformed and combined into very intricate and artificial systems. There is nothing in the loaf of bread, sliced and wrapped in wax paper to suggest the wheat field, the mill, or even the oven wherein the dough was baked. The man of an earlier generation traveling by foot or horse was aware of the mechanism which carried him and of the nature of the terrain which he traversed. Landmarks, springs and trails were a necessity. Today he moves, over roads for which engineers have changed the face of nature, in a mechanism whose operation is to him often a sealed book and at speeds the human body in a former generation could attain only by leaping down from some high cliff. He may experience infinite difficulty with his automobile because of the failure of a tiny copper wire hidden from his sight and put in place by a mechanic he never saw. Whereas his grandfather was obliged to see that the horse which carried him was fed hay, grass, or grain and given fresh water, the man of today drives up to a station and, in exchange for stamped metal or printed paper, has another transfer a measured amount of explosive fluid into the tank of his car. This fluid, instead of being a visible product of the fields by the roadside, like the grass and the grain which fed his grandfather's horse, has been obtained from perhaps a mile below the earth's surface, refined and transported to the station by the coördinated efforts of a large number of highly specialized workers. These workers have been guided at every possible step by practices based upon principles derived through the deliberate, systematic study of nature.

One might easily draw a similar comparison between personal combat, or even the wars of a few decades ago, and modern mechanized war. The story of modern soldiers who have fought and been wounded in several major engagements without seeing an enemy may be trite, but it is none the less significant. The modern high command is extremely remote from the rank and file. Direct personal leadership is a task for those far below them. The general officer who exposed himself to fire in the Philippines, sharing the danger of his men, was censured quite as much as he was mourned. Technical competency, the ability to coördinate an intricate mechanism is more the business of a great general than is personal courage, assuming he can reach high rank without the latter trait. And to describe the rôle of the individual soldier one has to borrow a term from the machine age itself—he is a tooth on a cogwheel in a vast impersonal mechanism.

The analogy applies to the individual in peace as in war. For in addition to insulating man against contact with his physical environment, modern scientific technology has produced a specialization of work much more minute than the world has ever seen. The continuity of achievement known to the craftsman is gone. A group of carefree students watching a man paste colored labels on new brooms with lightning speed, were instantly sobered, even depressed, when they learned he had done nothing else for fifteen years! Yet compared to many modern workmen, he is a versatile genius. Whereas formerly the humble workman (so-called) had at least the privilege of seeing a thing through from start to finish, that privilege today is reserved for those who direct his efforts and—this is important—who have no physical contact with the materials which are being elaborated. The psychic and social results of this situation are of course profound.

It is evident that modern technology, whatever its merits, has exerted a dissolving influence, loosening the relation of the individual to the physical environment and destroying his sense of the continuity of his own efforts. Whatever his older ideas of relationship to the universe, his present ones are inevitably confused. The old bearings and points of reference are gone, and the consequences of that fact are important. Anything approaching equilibrium or adjustment is impossible. Impossible in the first place because the new environment is in itself too confused, too rapidly changing to permit the necessary adjustment. Impossible in the second place because of the resultant psychic bewilderment and disintegration, making still more hopeless any ordered approach to the entire problem.

The evidence of this is manifold. Examine the basis of technological development—power and its application. One finds no ordered, reasonable approach. So long as coal was the chief source, it was tapped and exploited with no reference to economy, utility, permanence of supply, or ordinary human justice to the workers who mined it. Mines were opened up in full competition wherever they could be found, often to be gutted and ruined, alternately to flood the market and starve it. The oil industry which had coal as an example before it, has fared no better. Pools are opened up, sometimes we are assured, a decade in advance of any justifiable need, exploited in extravagant fashion, and exhausted with great speed, often in the face of a patent oversupply. The resources of both coal and oil are, we know, strictly limited, vast though they may be in aggregate. Now that coal enjoys a respite, we are assured that when the oil is gone, we can return to it and by proper chemical processes, secure what we need in the way of motor fuel and other hydrocarbons. Beyond that there is nothing but the possibility of wider use of water power, not always conveniently located, and an unbounded faith in future scientific miracles.

The situation is no different with respect to any other natural resource the environment affords, whether it be used for power or material. Forests, lead and zinc, the soil itself have all had unleashed against them the diabolical power of modern technology, contrived and planned with the utmost skill of the human mind—but *with no plan behind the plan*. The net effect to date has been primarily to accelerate the speed with which these things are consumed, thus hastening the day when mankind will be obliged to go on short rations, or devise new substitutes.

What justifies these really ruthless measures, of course, is the assumption that they create goods for the wider enjoyment among human beings, the measure of such benefit being the willingness of the people to purchase. Yet one does not have to examine modern culture long to learn that the desire to purchase seems to run ahead of the ability to do so, and at times behind the capacity of modern technology to produce. An additional technique has been developed, combining empirical with some scientific knowledge of the human mind—advertising as a means of causing consumption to catch up with production, or the capacity to produce.

Primarily at least advertising does not concern itself with any adjustment between purchase and the capacity to purchase—the sole function is to create desire. And upon this desire, so cleverly created, the only check is the mechanism of credit, which periodically shows itself unequal to the strain. This system of *manufacture—advertising—credit—consumption—adjustment* describes our present technological culture mechanism quite as accurately as the term "foodgatherer" or "pastoral nomad" does a primitive society.

One of the vital links in this chain is credit. Credit, under our system, is vital to the manufacturer, advertiser, and consumer. Credit can check or speed up the forces making for rapid change. This it is easy to demonstrate. Bankers, not operators, for a long time delayed the modernization of railroad rolling stock; if a railroad happened to become bankrupt, it was a great convenience to have cars and engines precisely like those used on every other road in the United States, so that the financiers who held the sack could dispose of them without difficulty. And so experiment and improvement was discouraged, until there was an overwhelming need for it. If you wish to travel on one of the new streamlined trains, it is well to make reservations some time ahead, so cautious and gradual has been the change.

Without credit the wisest measures may fail; by means of credit the power of a fool or knave is multiplied. It is not too much to say that one of the most potent instruments for adjusting and coördinating the modern culture pattern is credit—a thing as invisible as the electric power whose use it makes possible.

Even so, let us remember that credit is but an instrumentality, no matter how powerful it may be. Back of it lies the question of intent, will, direction. Advertising, as we shall see, is a most effective means of modifying the intent of the consumer. Yet it does not bring us to the bottom of the problem. Is it possible to have some greater intention behind all of the activities of the human community? Is it necessary? How can it be created and guided?

We have seen how modern man has lost his fear of and immediate, responsible contact with nature. Yet we must remember that he still works within the frame of natural cause and effect. He is as responsible to nature as he ever was. Moreover his scientific approach has created a mechanism of living which is, in itself, a source of new problems. Yet it has given him also the means of understanding. He need not, indeed, dare not, avoid the problem of a wise use and coördination of the forces he has invoked.

Chapter 18

THE CHALLENGE

PRINTING by movable type, world-wide exploration, and ultimately the invention of rapid transport and almost instant communication have broken down the ancient, purely physical barriers to the dream of human brotherhood. Sad to say, these developments, along with many devices of our high technology in the hands of huge masses of conscripted manpower, are still being used in the destruction of resources, people, and cultures.

Economists, political scientists, psychologists, religious leaders, and others may have their own varied explanations of this ghastly process, but it still goes on. Thanks to human ingenuity and fossil power, a "silent majority" in more fortunate lands lives in relative comfort, unprovoked to effective action even by a highway death-toll (much of it due to alcohol) that exceeds our losses in battle.

We are bemused by a degree of leisure conferred upon us by machines using fossil energy accumulated during the ages instead of the muscles of men and animals fueled by food produced in brow-sweat. Mass production is being employed by increasingly large and impersonal organizations to turn out consumer goods faster than they can be utilized. Persuasion becomes a greater problem than production, creating a constant din of temptation. For the favored ones already confused about how to employ their new leisure, this adds further uncertainty about how they will spend their money.

Now the justification urged for a mechanized society is the saving of human effort. But in corporate accounting it also saves the cost of labor, despite the fact that many are left jobless to be supported by taxes from the profits of industry and wages of those who do have work. Granting the dubious premise that the unemployed are thus taken care of, the real cost to society is the growing number who are without a significant role in it. This is the obverse side of the new leisure, by no means a blessing.

What the machine has done is to intensify an old condition, visible wherever, as in the Orient, the ratio between human numbers and useful work passes the danger point. There is a stage in population increase where it is beneficial, thanks to the need for a diversity of functions, specialists ministering to the welfare of the group. But this stage is soon exceeded, as records of ancient Egypt eloquently prove. There are today too many people, not primarily because they crowd each other or cannot conceivably be fed, but because too many lack any contributing function.

The earth and its resources, though vast, are finite. We have been using these resources (including space) without much control, in ways which spoil the environment and lessen its power to sustain life. This process can be reversed. Good husbandry could recycle plant nutrients. Metals, now often dissipated beyond recovery, could be salvaged for reuse. Mineral fuels, coal, oil, and gas, once converted into other forms of energy, are lost; they should be used providently, with least waste, and in ways that do not contaminate the environment as at present.

Space, if we consider quality as well as area, is Earth's basic resource. It cannot, as populations have done, increase with the passing of time. For my address as retiring president of the American Association for the Advancement of Science in 1957, I used as title "The Inexorable Problem of Space," noting that in the minds of many the word had come to connote outer rather than terrestrial space.[1] Yet the destiny of our race cannot be separated from our management of the latter. My plea was for the highest priority to the space in which we live.

Such a proposal can easily be misunderstood, and was. Grapevine reported that an article featuring it was prepared by the staff of a widely circulated journal, but killed on order from above. I had not questioned the magnificent technical and personal achievements, later to be so spectacularly climaxed by moon landings, nor the right of scientists to learn all they could about the universe.

What concerned me, as it still does, was the relative amount of treasure, brains, and effort being given to exploring outer space while our immediate environment was festering under exploitation and failure to use knowledge we already had. The same criticism—quite apart from humanitarian issues—holds for armed conflict, most of all when it is preventable by statesmanship.

There is no longer much need to catalog the ills we have brought to our environment or their threat to our

[1] *Science*, 127:9–16 (January 3, 1958).

future. Daily press and periodical literature furnish a running account of the pollution of air and water, urban decay, population pressure, hunger, and malnutrition, as well as the threat of impending shortages. There is also a growing literature of protest, at times intemperate, now and then inaccurate—ammunition for those ready to label the whole environmental movement an outburst of emotionalism.

Aside from the fact that little is accomplished in human affairs without an emotional drive behind it, neither side of the argument has any monopoly on emotion. To see it in full flower, one need only witness the meeting of a board whose interests are threatened. Yet in fairness I can testify from personal experience that the best of business and industrial leadership is far from indifferent to environmental problems; often it has been helpful in dealing with them.

As enterprise—for example timber production—has become a corporate rather than an individual operation, management tends to be concerned with continuing solvency, a sound general economy, and public good will. To say this is not to deny the existence of indifference, stupidity, rascality, or misguided enthusiasm. Important as it may be to recognize them, it is still more important not to imagine them where they are not present. The good-guy, bad-guy picture of society is neither true nor constructive.

A fellow passenger who had lived in the Near East had this to say of Nasser: "We go in to complain to him of some of his actions. He patiently tells us, 'This is what I'm up against. What would you do?' And I must say we leave with the feeling that he is a pretty reasonable fellow."

It would hardly do to muzzle any critic who cannot suggest a workable remedy for the evil he points out; yet his criticism is a lot more useful if it includes alternatives of a feasible sort. Protest against a dam that would destroy scenic or other values should include scrutiny of the ends to be served by such a dam; if these ends seem legitimate, plans for meeting them otherwise should be an integral part of the protest.

Included in the restlessness of modern youth is a strong concern for better environment, along with an impatience for results. There is also a feeling that not much can be accomplished within the present framework of society, but how prevalent this mood may be is only a statistical guess. Whatever the degree of rebellion, youth is entitled to know the resources of our present system for bringing about reforms. And from science (i.e., scientists) it has a right to a sense of direction so far as creating an ecologically viable system is concerned.

The Utopian challenge, long an exercise of the literary imagination, is today a grim reality. But Utopian writing, whatever its charm, has been weak on ways and means in terms of social process. And while attentive to humane, not to say idyllic, relationships necessary to decent and stable community life, such writing has ignored the compulsive physical conditions that must be met in any hope of permanency and order.

In planning a journey, one needs to know where he is heading, how he will get there, and what, once arrived, he will do. This homely reminder will serve us conveniently in what follows. We need only assume that the human adventure on this planet is worth continuing for as long as possible and in a way that deserves to continue.

Life has existed on our planet for above two billion years by virtue of what has been called the balance of nature. In terms of physical science this has been an open, steady state —a system in which energy from the sun has been used not only to build living communities but to keep them in repair. The materials needed to sustain life have not been wasted or locked up, but—thanks to the diversity of living organisms—kept available for use and reuse by succeeding generations. In numbers, behavior, and mutual relationships, all surviving plants and animals have been obliged to conform to the nature of their physical environments or be eliminated.

It takes little imagination to see how far our neotechnical society of the past century has fallen short of meeting these conditions. The direction in which it must move, for the stark purpose of self-preservation, is nowhere better stated than in a brief editorial in *Science* for November 14, 1969, by René Dubos, entitled "A Social Design for Science." To move toward this goal, our numbers must be controlled, waste eliminated, energy used with judgment, and materials recycled.

A further hint from natural models has been questioned, but the burden of proof rests on the questioners. This hint is that stability and persistence rest upon diversity rather than uniformity. Applied to human societies, this would mean encouragement of ethnic, cultural, and functional differences within societies in the interest of permanence. Our disregard of this principle shows not only in racism but in the prevalent asinine belief that both employability and social respectability depend upon the possession of a college degree. This is at once an affront to a host of self-educated, often eminent, individuals, and a source of badly crowded, troubled institutions of higher learning.

Even though we come to accept the necessity of working toward a steady state as the price of survival, the process of change will be tough. We have not yet learned to tolerate, let alone encourage, diversity; this, despite the obvious revolt against social and economic conformity by youth from homes of privilege. At least the question of justice to ethnic minorities is very much before the house and not likely to be tabled before it is resolved.

Any move in the direction of ecological equilibrium is bound to bring under scrutiny our economic system and the philosophy that sanctions it. Attempts to set up, control, or modify economic systems are an old story. Spartan coins were unwieldy to discourage attempts to accumulate any more than necessary. Rental for the use of money, which we call interest, was considered immoral by the church until medieval times. Free private enterprise has never had universal blessing, while socialism and the communism often confused with it are protests against the system of *laissez-faire*.

From where I stand, it seems as though these various systems and doctrines have developed (or been set up) as ways to get on with business while managing the distribution and enjoyment of wealth among individuals or groups. What seems to have been lacking is precisely what is involved in moving from our present curious mixture of free and managed economy to an open steady state—that is, protection of the environment as the physical source of wealth. One could, of course, transpose into the more familiar setting of the distribution of wealth by pointing out that the rights of future generations are being insured by creating sound ecological conditions.

However we may try to justify the changes which are ecologically necessary—whether as basic to a sound permanent economy, essential to human survival, required by social justice to the future, or as the road to ethical and esthetic decency—such changes will not be painless. The obstacles even to an innocuous degree of urban and regional planning should tell us this. Whether we try to change the rules or merely to enforce those we have, we run into conflicts of interest.

Having seen Florida, Arizona, California, and Hawaii (notably but not exclusively) destroy much that has made them attractive, I am only moderately optimistic. But I am cheered by a growing tendency to realize that improvement is more than a matter of curbing the misuse of power. The individual, we are reminded, needs to do some curbing on his own account.

For example, the use of electric power has multiplied much faster than population—four times as fast in New England, in fact. Concerned youth is asking, "Is this necessary?" Two generations ago a family saving up for its first automobile used the slogan "Tight with the light." As a biologist knowing that human muscles must be used to keep the body healthy, I have a jaundiced view of the host of cute devices that use electric current to save effort and time—for what?

The need for self-restraint suggests a degree of social discipline new to our society and complicated in several ways. Balancing overconsumption by the privileged are the needs for food, clothing, shelter, and useful work by a considerable portion of the American public. There is growing insistence, best evidenced in the demand for a guaranteed annual income, that these elementary needs be met.

It is now also clear that the citizen will have to pay heavily for a better environment, in both taxes and prices. Separation of sanitary from storm sewers and better waste treatment is costly business. Sewer reconstruction in one city of less than half a million that now drains into Lake Erie will require an outlay of eighty million dollars—an amount not now available.

Industry is already feeling the pressure to review operations and products that pollute air and water. This again is a matter of big money. Pratt-Whitney spent one million dollars to insure safe effluent from a new plant. Praise from the public was balanced by censure from other industries that could not afford to follow such an expensive example.

Costs to industry will have to be met either by increased prices or in some cases by subsidy from a public which after all has long tolerated, if not invited, the use of water and air as free sewers for industrial wastes. The normal response of the individual faced with this situation is to demand higher wages or cut down on his buying wherever he can. Neither course offers a rosy prospect to corporate profits under our present system. Nor does a more restrained and disciplined pattern of consumption to prevent needless waste of natural resources have a strong public appeal.

To appreciate how deeply such changes must run against the American grain, one should read Mrs. Trollope's account of her visit to this country in 1827-32. We can pass over her revulsion at tobacco chewing and indiscriminate spitting—a trait not curbed by law for eighty years and then on hygienic rather than esthetic grounds. More serious was her observation on the universal pursuit of the nimble dollar, *regardless of consequences*. I should add that she admired the natural beauty of our country and that Mark Twain, responding to the abuse heaped on her, declared that she wrote nothing but the truth.

Granting all the improvement that has taken place in more than a century and a half, our pocketbooks are still like sensitive nerve endings. Nor have the immense vistas of profit opened up by applied science improved matters. Not much monitoring of broadcasts or journals is needed to show that whatever becomes technically possible and profitable seems, if not ethically justified, at least obscured from ethical criticism.

Many conditions unfavorable to the health of human beings and their environment can be traced to practices that cater to the consumer and contribute to high industrial profits. To remedy these conditions will be expensive both for the consumer and for those who supply him. Neither will like this, but speedier resistance must come from management than from the public, since management is professionally sensitive to the reading of the balance sheet.

It is later than we think, although I do not share the vision of total disaster for humanity within the next few decades. We are a sturdy and resourceful species. Harrison Brown is probably correct in his view that if massive calamity strikes, it will be the agrarian rather than the highly urbanized and industrialized cultures that are likely to survive. But in any event we must begin to accept and plan for profound changes in our present values and beliefs as the price of survival. Preserving at all costs our defense of human dignity and the right of all to a good life, we must reexamine our priorities to preserve what is really essential.

How this is to be accomplished is by no means clear. Human motives are a variable mixture of necessity, conviction, and expediency. The late Hendrick van Loon in his *Geography* years ago reminded us that we are all fellow-passengers on a lonely planet—a message echoed by Adlai Stevenson's "Space-Ship Earth" and dramatically emphasized by our moon shots. This is what we have and must begin to accept as a working principle, getting rid of our belligerent and defensive attitude toward other ways of life than our own, as though we had everything to teach and nothing to learn.

We desperately need to keep in mind the ancient precept that man does not live by bread alone, translating bread into those symbols of wealth we call dollars. To another ancient injunction, that it doth not profit to gain the whole world and lose our soul, we add, and lose the world through coveting it. What price indeed for an ever growing Gross National Product? True growth leads to order; any other is pathological.

Born and reared in the tradition of conservatism, I had the privilege to know Norman Thomas not long before his death. Many of the things he had advocated are now accepted as a matter of course. One of his great satisfactions, after a long and often stormy career, was the certainty that he would come to be recognized as one of the truest conservatives of his day. In his insistence that the economic system must enhance, not limit, the dignity and brotherhood of man and be based on reverence for the environment, he surely pointed the way we must follow to be saved from ourselves.

So much for principles. The insistent plea "Yes, but what can I do?" calls for practical suggestions. Most of the effective workers for a better environment that I have known through the years have early learned to see and appreciate nature—not as hobbyists or doctrinaires, but as understanding observers of the world in which we live. In most cases they owe this gift to individual teachers, to parents, or (as did Darwin) to companions. No education, however elaborate, that neglects this feature deserves the name.

Three instances come to mind. Miss Melrose, a science supervisor in Cleveland, by creating an interest in nature